"十三五"国家重点出版物出版规划项目
面向可持续发展的土建类工程教育丛书

土木工程专业英语

鲁 正 编

机械工业出版社

本书的编写基于同济大学近十几年对土木工程专业英语课程的教学探索和大量的素材积累；全书涵盖土木工程各个领域的基本英语用法，有点有面；紧跟目前科研和工程热点，比如 BIM 技术、住宅工业化技术等；实用性很强，最后一章讲述了科技论文写作。全书分为 19 个单元，并将其归为 5 大部分，分别为结构工程、土木工程其他分支、土木工程新技术、项目管理和案例分析、专业英语写作。

本书可作为土木工程专业及相关专业的教材，也可供专业技术人员了解专业知识、提高英语水平时使用。

本书配有授课 PPT 和 Exercises 部分的参考答案等资源，免费提供给选用本书的授课教师，需要者请登录机械工业出版社教育服务网（www.cmpedu.com）注册下载。

图书在版编目（CIP）数据

土木工程专业英语/鲁正编. —北京：机械工业出版社，2017.12
（2023.8 重印）
（面向可持续发展的土建类工程教育丛书）
"十三五"国家重点出版物出版规划项目
ISBN 978-7-111-58559-6

Ⅰ. ①土… Ⅱ. ①鲁… Ⅲ. ①土木工程-英语-高等学校-教材 Ⅳ. ①TU

中国版本图书馆 CIP 数据核字（2017）第 292092 号

机械工业出版社（北京市百万庄大街 22 号　邮政编码 100037）
策划编辑：李　帅　责任编辑：李　帅　臧程程　责任校对：高亚苗
封面设计：张　静　责任印制：单爱军
北京虎彩文化传播有限公司印刷
2023 年 8 月第 1 版第 6 次印刷
184mm×260mm · 19.75 印张 · 479 千字
标准书号：ISBN 978-7-111-58559-6
定价：49.00 元

电话服务　　　　　　　　　网络服务
客服电话：010-88361066　　机 工 官 网：www.cmpbook.com
　　　　　010-88379833　　机 工 官 博：weibo.com/cmp1952
　　　　　010-68326294　　金 书 网：www.golden-book.com
封底无防伪标均为盗版　　　机工教育服务网：www.cmpedu.com

前　言

随着"高铁外交"与中国"一带一路"倡议的实施，我国的土木工程在国际舞台上发挥着越来越重要的作用，涉外工程、科研项目越来越多，国际化趋势持续加速。国内土木工程领域的众多龙头公司已将开拓海外市场作为可持续发展的长期战略并取得了卓越成效。例如我国建筑行业的领军企业——中国建筑工程总公司（简称"中国建筑"），在2016年度新签海外合约额1264亿元，同比增长13%；实现海外营收796亿元，同比增长30%，首次突破百亿美元大关。中国交建、中国铁建、中国中车等国内著名的国际工程承包商在海外的订单额和营收也以每年20%左右的增速高速发展。

繁荣景象的延续需要后备力量的长久支持，土木工程专业英语的教学为国家"一带一路"倡议提供可靠人才保障。我国对外工程的长足发展不能仅依靠成本优势，更要凭借高端竞争优势，而企业竞争力的核心便是"人才"。目前，我国土木工程领域中既能熟练使用专业英语，又有过硬技术本领的人才极度匮乏。党的二十大报告指出："我们要坚持教育优先发展、科技自立自强、人才引领驱动，加快建设教育强国、科技强国、人才强国，坚持为党育人、为国育才，全面提高人才自主培养质量，着力造就拔尖创新人才，聚天下英才而用之。"因此，人才的培养应紧追企业海外拓展的步伐，打破复合型双语人才匮乏的瓶颈，培养出与国际接轨、具有核心竞争力的人才是当务之急。

另一方面，我国高校毕业生就业形势异常严峻，土木工程专业英语的学习为毕业生的顺利就业增添筹码。目前，我国每年应届毕业生700多万人并逐年增加，众多毕业生为找工作费尽周折。随着企业工程的进一步国际化，精通专业英语的高质量复合型人才必定受到国内外著名企业的追捧。

土木工程专业英语的教学利于国家战略、利于学生发展，培养精通土木工程专业英语的高层次复合人才已成为广大土木工程专业相关管理者和一线教师的共识。

本书选择了19篇专业英语阅读材料，包括结构工程、土木工程其他分支、土木工程新技术、项目管理和案例分析以及专业英语写作等方面的内容。学生通过学习土木工程专业英语这门课程可以掌握英文科技专业书刊中各种句型的表达方式、语法、主要专业词汇及写作技巧，从而为今后阅读英文科技文献、撰写英文科技论文、从事涉外施工和涉外设计等工作奠定基础。

本书内容来源于同济大学土木工程学院结构工程与防灾研究所多年积累的教学讲义。讲义的第1版由蒋欢军和吕西林于2000年选编，第二版由周颖于2007年选编，第3版由鲁正于2015年选编。经过17年的教学实践和积累，最终在2017年12月，由鲁正选编形成本书。由于时间和水平关系，书中不妥之处在所难免，敬请批评指正。

编　者

目 录

前言

Part 1　Introduction to Structural Engineering

Unit 1　Introduction to Reinforced Concrete Design ·· 2
1.1　Concrete, Reinforced Concrete, and Prestressed Concrete ····················· 2
1.2　Structural Forms ·· 4
1.3　Loads ·· 7
1.4　Serviceability, Strength, and Structural Safety ·· 11
1.5　Design Basis ·· 14

Unit 2　Introduction to Prestressed Concrete ·· 19
2.1　Introduction ·· 19
2.2　Effects of Prestressing ··· 20
2.3　Sources of Prestress Force ··· 24
2.4　Prestressing Steels ·· 27
2.5　Concrete for Prestressed Construction ·· 28

Unit 3　Introduction to Steel Structures ·· 33
3.1　Structural Design ··· 33
3.2　Principles of Design ··· 33
3.3　Historical Background of Steel Structures ·· 34
3.4　Loads ·· 35
3.5　Types of Structural Steel Members ·· 43

Unit 4　Seismic Design ·· 51
4.1　Introduction ·· 51
4.2　Structural Response ·· 52
4.3　Seismic Loading Criteria ·· 56

Unit 5　Composite Construction ·· 64
5.1　Overview ··· 64
5.2　Pre-stressed Concrete Composite Slabs ··· 67

Unit 6　Introduction to Foundation Analysis and Design ···································· 73
6.1　Foundations—Definition and Purpose ··· 73

6.2　Foundation Classifications ·· 74
6.3　Foundation Site and System Economics ··· 75
6.4　General Requirements of Foundations ·· 77
6.5　Foundation Selection ··· 77
6.6　SI and Fps Units ··· 78
6.7　Computational Accuracy Versus Design Precision ···································· 79

Part 2　Introduction to Other Branches of Civil Engineering

Unit 7　Introduction to Bridge Engineering ··· 84
7.1　Reinforced Concrete Girder Bridges ··· 84
7.2　Arch Bridges ·· 85
7.3　Steel Bridges ·· 86
7.4　Truss Bridges ··· 86
7.5　Plate and Box Girder Bridges ·· 87
7.6　Cable Stayed Bridges ·· 88
7.7　Suspension Bridges ·· 89

Unit 8　Introduction to Underground Engineering ···································· 94
8.1　The Future of Underground Infrastructure in Holland ······························· 94
8.2　Seismic Design and Analysis of Underground Structure ····························· 98
8.3　Performance of Underground Facilities During Seismic Events ···················· 99

Unit 9　Introduction to Traffic Engineering ··· 108
9.1　Introduction ·· 108
9.2　Traffic Management and Control ·· 112

Unit 10　Hydraulic Engineering ·· 119
10.1　Introduction ·· 119
10.2　Types of Hydraulic Structures ··· 119
10.3　Layout of Hydraulic Projects ··· 121
10.4　Classification of Hydraulic Projects and Their Design Safety Standards ······ 124
10.5　Water Resources and Hydropower Resources in China ··························· 125
10.6　Hydraulic Engineering in China ··· 126
10.7　Purposes of Hydraulic Projects ··· 127

Part 3　New Technology in Civil Engineering

Unit 11　Industrialized Housing Systems Construction in China ················· 136
11.1　Introduction to Industrialized Housing Systems ···································· 136
11.2　Types of Industrialized Housing Systems (IHS) ···································· 138
11.3　Selecting Housing Industrialization Grading Indicators ··························· 139
11.4　Basic Methodology of Hierarchical Clustering Method ··························· 141
11.5　Region-Based Housing Industrialization Grading Analysis ······················ 143

11.6	Development Bottlenecks of Housing Industrialization in China	145
11.7	Advantages and Disadvantages of Industrialized Housing Systems	145
11.8	Key Implementations to Housing Industrialization	146

Unit 12　Passive Base Isolation with Merits and Demerits Analysis …… 150

12.1	Introduction	150
12.2	Concept of Base Isolation	152
12.3	Base Isolation Systems	154
12.4	Merits and Demerits Analysis	162

Unit 13　Supplemental Energy Dissipation: State-of-the-art and State-of-the-practice …… 167

13.1	Introduction	167
13.2	Basic Principles	169
13.3	Passive Energy Dissipation	171
13.4	Active, Hybrid and Semi-active Control Systems	176
13.5	Concluding Remarks	186

Unit 14　Introduction to 3D Printing of Buildings and Building Components …… 190

14.1	3D Printing Technology and Materials	190
14.2	Examples of 3D Printing Building	191
14.3	Application of 3D Printing Reproduction of Historical Building Ornamental Components	194
14.4	Preparation of Computer Models for 3D Printing	201

Unit 15　BIM: Acceptance Model in Construction Organizations …… 208

15.1	Introduction to BIM	208
15.2	Applications of Building Information Modeling	209
15.3	Role of BIM in the AEC Industry: Current and Future Trends	210
15.4	BIM Benefits: Case Studies	212
15.5	Return on Investment Analysis	218
15.6	BIM Risks	219
15.7	BIM Future Challenges	221

Part 4　Project Management and Case Study

Unit 16　Internationalization of Chinese Construction Enterprises …… 228

16.1	Abstract	228
16.2	Introduction	228
16.3	Construction Industry in China	229
16.4	Development of Chinese International Contractors	233
16.5	Engineering News Record Top 35 Chinese International Contractors	237
16.6	Literature Review	237

16.7	Analysis of Engineering News Record Top 35 Chinese International Contractors	240
16.8	Conclusion	243

Unit 17　Project Management and Administration … 248

17.1	The Need for Project Management	248
17.2	Project Organization	248
17.3	The Project Manager	249
17.4	The Project Superintendent	250
17.5	Jobsite Computers	251
17.6	Aspects of Project Management	251
17.7	Field Productivity	252
17.8	Project Administration	253
17.9	Project Meetings	253
17.10	Schedule of Owner Payments	254
17.11	Shop Drawings	254
17.12	Approval of the Shop Drawings	255
17.13	Quality Control	256
17.14	Expediting	257
17.15	Deliveries	259
17.16	Receiving	259
17.17	Inspection of Materials	260
17.18	Subcontractor Scheduling	261
17.19	Record Drawings	262
17.20	Disbursement Controls	262
17.21	Job Records	263
17.22	The Daily Job Log	263
17.23	Claims	264

Unit 18　Design and Construction of the Jin Mao Tower … 269

18.1	The structure system	269
18.2	Foundation engineering	272
18.3	Wind engineering	273
18.4	Earthquake engineering	274
18.5	Unique structural engineering solutions/On-site structure monitoring	275
18.6	Conclusions	277

Part 5　Scientific English Writing Skills

Unit 19　Scientific English Writing Skills … 284

19.1	科技英语的基本特点	284
19.2	科技论文的组成	287

参考文献 … 304

Part 1

Introduction to Structural Engineering

Unit 1

Introduction to Reinforced Concrete Design

1.1 Concrete, Reinforced Concrete, and Prestressed Concrete

Concrete is a stonelike material obtained by permitting a carefully proportioned mixture of cement, sand and gravel or other aggregate, and water to harden in forms of the shape and dimensions of the desired structure. The bulk of the material consists of fine and coarse aggregate. Cement and water interact chemically to bind the aggregate particles into a solid mass. Additional water, over and above that needed for this chemical reaction, is necessary to give the mixture the workability that enables it to fill the forms and surround the embedded reinforcing steel prior to hardening.[1] Concretes with a wide range of properties can be obtained by appropriate adjustment of the proportions of the constituent materials. Special cements (such as high early strength cements), special aggregates (such as various lightweight or heavyweight aggregates), admixtures (such as plasticizers, air-entraining agents, silica fume, and fly ash), and special curing methods (such as steam-curing) permit an even wider variety of properties to be obtained.

混凝土特性

These properties depend to a very substantial degree on the proportions of the mix, on the thoroughness with which the various constituents are intermixed, and on the conditions of humidity and temperature in which the mix is maintained from the moment it is placed in the forms until it is fully hardened.[2] The process of controlling conditions after placement is known as *curing*. To protect against the unintentional production of substandard concrete, a high degree of skillful control and supervision is necessary throughout the process, from the proportioning by weight of the individual components, through mixing and placing, until the completion of curing.

The factors that make concrete a universal building material are so pronounced that it has been used, in more primitive kinds and ways than at present, for thousands of years, starting with lime mortars from 12000 to 6000 BCE in Crete, Cyprus, Greece, and the Middle East. The facility with which, while plastic, it can be deposited and made to fill forms or molds of almost any practical shape is one of these factors.[3] Its high fire and weather resistance is an evident advantage. Most of the constituent materials, with the exception of cement and additives, are usually available at low cost locally or at small distances from the construction site. Its compressive strength, like that of natural stones, is high, which makes it suitable for members primarily subject to compression, such as columns and arches. On the other hand, again as in natural stones, it is a relatively brittle material whose tensile strength is small compared with its compressive strength. This prevents its economical use in structural members that are subject to tension either entirely (such as in tie-rods) or over part of their cross sections (such as in beams or other flexural members).

To offset this limitation, it was found possible, in the second half of the nineteenth century, to use steel with its high tensile strength to reinforce concrete, chiefly in those places where its low tensile strength would limit the carrying capacity of the member. The reinforcement, usually round steel rods with appropriate surface deformations to provide interlocking, is placed in the forms in advance of the concrete. When completely surrounded by the hardened concrete mass, it forms an integral part of the member. The resulting combination of two materials, known as reinforced concrete, combines many of the advantages of each: the relatively low cost, good weather and fire resistance, good compressive strength, and excellent formability of concrete and the high tensile strength and much greater ductility and toughness of steel. It is this combination that allows the almost unlimited range of uses and possibilities of reinforced concrete in the construction of buildings, bridges, dams, tanks, reservoirs, and a host of other structures.

In more recent times, it has been found possible to produce steels, at relatively low cost, whose yield strength is 3 to 4 times and more that of ordinary reinforcing steels. Likewise, it is possible to produce concrete 4 to 5 times as strong in compression as the more ordinary concretes. These high-strength materials offer many advantages, including smaller member cross sections, reduced dead load, and longer spans. However, there are limits to the strengths of the constituent materials beyond which certain problems arise. To be sure, the strength of such a member would increase roughly in proportion to those of the materials. However, the high strains that result from the high stresses that would otherwise be permissible would lead to large deformations and consequently large deflections of such members under ordinary loading conditions. Equally important, the large strains in such high-strength reinforcing steel would induce large cracks in the surrounding low tensile strength concrete, cracks that not only would be unsightly but also could significantly reduce the durability of the structure. This limits the useful yield strength of high-strength reinforcing steel to 80 ksi (1ksi = 6.895MPa) according to many codes and specifications; 60ksi steel is most commonly used.

A special way has been found, however, to use steels and concretes of very high strength in combination. This type of construction is known as prestressed concrete. The steel, in the form of wires, strands, or bars, is embedded in the concrete under high tension that is held in equilibrium by compressive stresses in the concrete after hardening.[4] Because of this precompression, the concrete in a flexural member will crack on the tension side at a much larger load than when not so precompressed. Prestressing greatly reduces both the deflections and the tensile cracks at ordinary loads in such structures, and thereby enables these high-strength materials to be used effectively. Prestressed concrete has extended, to a very significant extent, the range of spans of structural concrete and the types of structures for which it is suited.

预应力混凝土

1.2 Structural Forms

The figures that follow show some of the principal structural forms of reinforced concrete. Pertinent design methods for many of them are discussed later in this volume.

Floor support systems for buildings include the monolithic slab-and-beam floor shown in Fig. 1-1, the one-way joist system of Fig. 1-2, and the flat plate floor, without beams or girders, shown in Fig. 1-3. The flat slab floor of Fig. 1-4, frequently used for more heavily loaded buildings such as warehouses, is similar to the flat plate floor, but makes use of increased slab thickness in the vicinity of the columns, as well as flared column tops, to reduce stresses and increase strength in the support region. The choice among these and other systems for floors and roofs depends upon functional requirements, loads, spans, and permissible member depths, as well as on cost and esthetic factors.

Fig. 1-1 One-way reinforced concrete floor slab with monolithic supporting beams (Portland Cement Association).

Fig. 1-2 One-way joist floor system, with closely spaced ribs supported by monolithic concrete beams; transverse ribs provide for lateral distribution of localized loads (Portland Cement Association).

Fig. 1-3 Flat plate floor slab, carried directly by columns without beams or girders (Portland Cement Association).

Fig. 1-4 Flat slab floor, without beams but with slab thickness increased at the columns and with flared column tops to provide for local stress concentration of forces.

Where long clear spans are required for roofs, concrete shells permit use of extremely thin surfaces, often thinner, relatively, than an eggshell. The folded plate roof of Fig. 1-5 is simple to form because it is composed of flat surfaces; such roofs have been employed for spans of 200 ft (1ft = 0.3048m) and more. The cylindrical shell of Fig. 1-6 is also relatively easy to form because it has only a single curvature; it is similar to the folded plate in its structural behavior and range of spans and loads. Shells of this type were once quite popular in the United Staler and remain popular in other parts of the world.

Fig. 1-5 Folded plate roof of 125 ft span, in addition to carrying ordinary roof loads, carries the second floor as well from a system of cable hangers; the ground floor is kept free of columns.

Fig. 1-6 Cylindrical shell roof providing column-free interior space.

Doubly curved shell surfaces may be generated by simple mathematical curves such as circular arcs, parabolas, and hyperbolas, or they may be composed of complex combinations of shapes. The hyperbolic paraboloid shape, defined by a concave downward parabola moving along a concave upward parabolic path, has been widely used, It has the interesting property that the doubly curved surface contains two systems of straight-line generators, permitting straight-form lumber to be used. The complex dome of Fig. 1-7, which provides shelter for performing arts events, consists essentially of a circular dome but includes monolithic upwardly curved edge surfaces to provide stiffening and strengthening in that critical region.

Bridge design has provided the opportunity for some of the most challenging and creative applications of structural engineering. The award-winning Napoleon Bonaparte Broward Bridge, shown in Fig. 1-8, is a six-lane, cable-stayed structure that spans St. John's River at Dame Point, Jacksonville, Florida. Its 1300 ft center span is the second longest of its type in the western hemisphere. Fig. 1-9 shows the Bennett Bay Centennial Bridge, a four-span continuous, segmentally cast-in-place box girder structure. Special attention was given to esthetics in this award-winning design. The spectacular Natchez Trace Parkway Bridge in Fig. 1-10, a two-span arch structure using hollow precast concrete elements, carries a two-lane highway 155 ft above the valley floor. This structure has won many honors, including awards from the American Society of Civil Engineers and the National Endowment for the Arts.

Fig. 1-7 Spherical shell in Lausanne, Switzerland. Upwardly curved edges provide stiffening for the central dome.

Fig. 1-8 Napoleon Bonaparte Broward Bridge, with a 1300 ft center span at Dame Point, Jacksonville. Florida (HNTB Corporation, Kansas City, Missouri).

Fig. 1-9 Bennett Bay Centennial Bridge, Coeur d'Alene, Idaho, a four-span continuous concrete box girder structure of length 1730 ft (HNTB Corporation, Kansas City, Missouri).

Fig. 1-10 Natchez Trace Parkway Bridge near Franklin, Tennessee, an award-winning two-span concrete arch structure rising 155 ft above the valley floor.

Cylindrical concrete tanks are widely used for storage of water or in waste purification plants. The design shown in Fig. 1-11 is proof that a sanitary engineering facility can be esthetically pleasing as well as functional. Cylindrical tanks are often pre-stressed circumferentially to maintain compression in the concrete and eliminate the cracking that would otherwise result from internal pressures.

结构形式

Concrete structures may be designed to provide a wide array of surface textures colors, and structural forms. Fig. 1-12 shows a precast concrete building containing both color changed and architectural finishes.

The forms shown in Fig. 1-1 to Fig. 1-12 hardly constitute a complete inventory but are illustrative of the shapes appropriate to the properties of reinforced or prestressed concrete. They illustrate the adaptability of the material to a great variety of one-dimensional (beams, girders, columns), two-dimensional (slabs, arches, rigid frames), and three-dimensional (shells, tanks) structures and structural components. This variability allows the shape of the structure to be adapted to its function in an economical manner, and furnishes the architect and design engineer with a wide variety of possibilities for esthetically satisfying structural solutions.

Fig. 1-11 Circular concrete tanks used as a part of the wastewater purification facility at Howden, England (Northumbrian Water Authority with Luder and Jones, Architects).

Fig. 1-12 Concrete structures can be produced in a wide range of colors, finishes, and architectural detailing (Courtesy of Rocky Mountain Prestress Corp).

1.3 Loads

Loads that act on structures can be divided into three broad categories: dead loads, live loads, and environmental loads.

Dead loads are those that are constant in magnitude and fixed in location throughout the lifetime of the structure. Usually the major part of the dead load is the weight of the structure itself. This can be calculated with good accuracy from the design configuration, dimensions of the structure, and density of the material. For buildings, floor fill, finish floors, and plastered ceilings are usually included as dead loads, and an allowance is made for suspended loads such as piping and lighting fixtures. For bridges, dead loads may include wearing surfaces, sidewalks, and curbs, and an allowance is made for piping and other suspended loads.

Source: From *Minimum Design Loads for Buildings and Other Structures*. Used by permission of the American Society of Civil Engineers.

Live loads consist chiefly of occupancy loads in buildings and traffic loads on bridges. They may be either fully or partially in place or not present at all, and may also change in location. Their magnitude and distribution at any given time are uncertain, and even their maximum intensities throughout the lifetime of the structure are not known with precision. The minimum live loads for which the floors and roof of a building should be designed are usually specified in the building code that governs at the site of construction. Representative values of minimum live loads to be used in a wide variety of buildings are found in *Minimum Design Loads for Buildings and Other Structures*, a portion of which is reprinted in Table 1-1. The table gives uniformly distributed live loads for various types of occupancies; these include impact provisions where necessary. These loads are expected maxima and considerably exceed average values.

Table 1-1 Minimum uniformly distributed live loads

Occupancy or Use	Live Load/psf[①]	Occupancy or Use	Live Load/psf[①]
Apartments (see residential)		Dining room and restaurants	100
Access floor systems		Dwelling (see residential)	
Office use		Fire escapes	100
Computer use	100	On single-family dwellings only	40
Armories and drill rooms	150	Garages (passenger cars only)	40
Assembly areas and theaters		Trucks and buses[②]	
Fixed seats (listened to floor)	60	Grandstands (see stadium and arena bleachers)	
Lobbies	100	Gymnasiums, main floors and balconies[③]	100
Movable seats	100	Hospitals	
Platforms (assembly)	100	Operating rooms, laboratories	60
Stage floors	150	Patient rooms	40
Balconies (exterior)	100	Corridors above first floor	80
On one and two-family residences only, and not exceeding 100ft²	60	Hotels (see residential)	
		Libraries	
Bowling alleys, poolrooms, and similar recreational areas		Reading rooms	60
		Stack rooms[④]	150
Catwalks for maintenance access	40	Corridors above first floor	80
Corridors		Manufacturing	
First floor	100	Light	125
Other floors, same as occupancy served except as indicated		Heavy	250
		Marquees and canopies	75
Dance halls and ballrooms	100	Office buildings	
Decks (patio and roof) Same as area served, or for the type of occupancy accommodated		File and computer rooms shall be designed for heavier loads based on anticipated occupancy	
		Lobbies and first-floor corridors	100
Offices	50	Schools	
Corridors above first floor	80	Classrooms	40
Penal institutions		Corridors above first floor	80
Cell blocks	40	First-floor corridors	100
Corridors	100	Sidewalks, vehicular driveways, and yards subject to trucking	250
Residential			
Dwellings (one and two-family)		Stadiums and arenas	
Uninhabitable attics without storage	10	Bleachers	100
Uninhabitable attics with storage	20	Fixed seats (fastened to floor)	60
Habitable attics and sleeping areas	30	Stairs and exit ways	100

(Continued)

Occupancy or Use	Live Load/psf[1]	Occupancy or Use	Live Load/psf[1]
All other areas except stairs and balconies	40	One and two-family residences only	40
Hotels and multifamily houses		Storage areas above ceilings	20
Private rooms and corridors serving them	40	Storage warehouses (shall be designed for	
Public rooms and corridors serving them	100	heavier loads if required for anticipated storage)	
Reviewing stands, grandstands, and bleachers[5]		Light	125
Roofs		Heavy	250
Ordinary flat, pitched, and curved roofs	20	Stores	
Roofs used for promenade purposes	60	Retail	
Roofs used for roof gardens or assembly purpose	100	First floor	100
Roofs used for other special purposes.[6]		Upper floors	73
Awnings and canopies	5	Wholesale, all floors	125
Fabric construction supported by a		Walkways and elevated platforms	60
lightweight rigid skeleton structure[7]		(other than exitways)	
All other constructions	20	Yards and terraces, pedestrians	100

① Pounds per square foot, 1psf = 47.88Pa.
② Garages accommodating trucks and buses shall be designed in accordance with an approved method that contains provisions for truck and bus loadings.
③ In addition to the vertical live loads, the design shall include horizontal swaying forces applied to each row of seats as follows: 24 lb⊖ per linear ft of seat applied in the direction parallel to each row of seats and 10 lb per linear ft of seat applied in the direction perpendicular to each row of seats. The parallel and perpendicular horizontal swaying forces need not be applied simultaneously.
④ The loading applies to stack room floors that support nonmobile, double-faced library bookstacks subject to the following limitations: (a) the nominal bookstack unit height shall not exceed 90 in.⊖; (b) the nominal shelf depth shall not exceed 12 in. for each face; and (c) parallel rows of double-faced bookstacks shall be separated by aisles not less than 36 in. wide.
⑤ Other uniform loads in accordance with an approved method that contains provisions for truck loadings shall also be considered where appropriate.
⑥ Roofs used for other special purposes shall be designed for appropriate loads as approved by the authority having jurisdiction.
⑦ Nonreducible。

In addition to these uniformly distributed loads, it is recommended that, as an alternative to the uniform load, floors be designed to support safely certain concentrated loads if these produce a greater stress. For example, office floors are to be designed to carry a load of 2000lb distributed over an area 2.5ft square (6.25ft^2), to allow for the weight of a safe or other heavy equipments, and stair treads must safely support a 300lb load applied on the center of the tread. Certain reductions are often permitted in live loads for members supporting large areas, on the premise that it is not likely that the entire area would be fully loaded at one time.

Tabulated live loads cannot always be used. The type of occupancy should be considered and the probable loads computed as accurately as possible. Warehouses for heavy storage may be designed for loads as high as 500psf or more; unusually heavy operations in manufacturing buildings may require an increase in the 250psf value specified in Table 1-1; special provisions must be made

⊖ 1lb: 磅, 1lb = 4.45N。
⊖ in: 英寸, 1in = 2.54cm。

for all definitely located heavy concentrated loads.

Live loads for highway bridges are specified by the American Association of State Highway and Transportation Officials (AASHTO) in its *LRFD Bridge Design Specifications*. For railway bridges, the American Railway Engineering and Maintenance-of-Way Association (AREMA) has published the *Manual of Railway Engineering*, which specifies traffic loads.

Environmental loads consist mainly of snow loads, wind pressure and suction, earthquake loads (i.e., inertia forces caused by earthquake motions), soil pressures on subsurface portions of structures, loads from possible ponding of rainwater on flat surfaces, and forces caused by temperature differentials. Like live loads, environmental loads at any given time are uncertain in both magnitude and distribution. The book, *Minimum Design Loads for Buildings and Other Structures*, contains much information on environmental loads, which is often modified locally depending, for instance, on local climatic or seismic conditions.

Fig. 1-13, from the 1972 edition of *Minimum Design Loads for Buildings and Other Structures*, which gives snow loads for the continental United States and is included here for illustration only. The 2005 edition gives much more detailed information. In either case, specified values don't represent average values, but are expected to upper limits. A minimum roof load of 20 psf is often specified to provide for construction and repair loads and to ensure reasonable stiffness.

Much progress has been made in developing rational methods for predicting horizontal forces on structures due to wind and seismic action. The book summarizes current thinking regarding wind forces and has much information pertaining to earthquake loads as well.

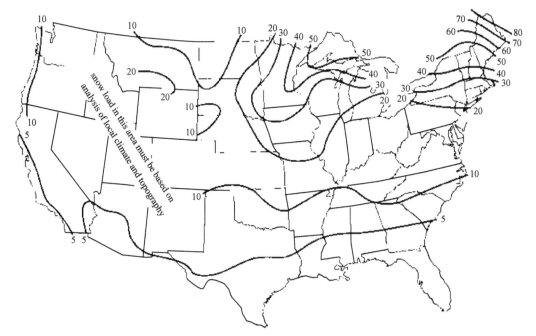

Fig. 1-13 Snow load in pounds per square foot (psf) **on the ground, 50-year mean recurrence interval.**

Wind pressures are specifically designed by per square foot of vertical wall surface. Depending upon locality, these equivalent static forces vary from about 10 to 50psf. Factors include basic wind speed, exposure (urban vs. open terrain, for example), height of the structure, the importance of the structure (i.e., consequences of failure), and gust effect factors to account for the fluctuating nature of the wind and its interaction with the structure.

Seismic forces may be found for a particular structure by elastic or inelastic dynamic analysis, considering expected ground accelerations and the mass, stiffness, and damping characteristics of the construction. However, the design is often based on equivalent static forces. The base shear is found by considering such factors as location, type of structure and its occupancy, total dead load, and the particular soil condition. The total lateral force is distributed to floors over the entire height of the structure in such a way as to approximate the distribution of forces obtained from a dynamic analysis.

■ 1.4 Serviceability, Strength, and Structural Safety

To serve its purpose, a structure must be safe against collapse and serviceable in use. Serviceability requires that deflections be adequately small; that cracks, if any, be kept to tolerable limits; that vibrations be minimized; etc. Safety requires that the strength of the structure be adequate for all loads that may foreseeably act on it. If the strength of a structure, built as designed, could be predicted accurately, and if the loads and their internal effects (moments, shears, axial forces) were known accurately, safety could be ensured by providing a carrying capacity just barely in excess of the known loads.[5] However, there are a number of sources of uncertainty in the analysis, design, and construction of reinforced concrete structures. These sources of uncertainty, which require a definite margin of safety, may be listed as follows:

1. Actual loads may differ from those assumed.
2. Actual loads may be distributed in a manner different from that assumed.
3. The assumptions and simplifications inherent in any analysis may result in calculated load effects—moments, shears, etc. —different from those that, in fact, act in the structure.
4. The actual structural behavior may differ from that assumed, owing to imperfect knowledge.
5. Actual member dimensions may differ from those specified.
6. Reinforcement may not be in its proper position.
7. Actual material strength may be different from that specified.

使用性、强度和结构安全

In addition, in the establishment of a safety specification, consideration must be given to the consequences of failure. In some cases, a failure would be merely an inconvenience. In other cases, loss of life and significant loss of property may be involved. A further consideration should be the nature of the failure, should it occur. A gradual failure with ample warning permitting remedial measures is preferable to a sudden, unexpected collapse.

It is evident that the selection of an appropriate margin of safety is not a simple matter. However, progress has been made toward rational safety provisions in design codes.

1.4.1 Variability of Loads

Since the maximum load that will occur during the life of a structure is uncertain, it can be considered a random variable. In spite of this uncertainty, the engineer must provide an adequate structure. A probability model for the maximum load can be devised by means of a probability density function for loads, as represented by the frequency curve of Fig. 1-14a. The exact form of this distribution curve, for any particular type of loading such as office loads, can be determined only on the basis of statistical data obtained from large-scale load surveys.[6] A number of such surveys have been completed. For types of loads for which such data are scarce, fairly reliable information can be obtained from experience, observation, and judgment.

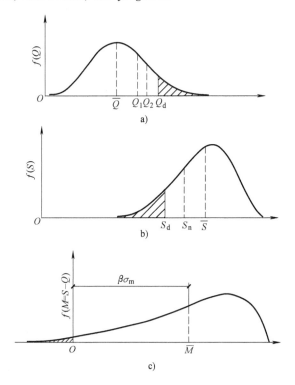

Fig. 1-14 Frequency curves for loads Q, strength S, and safety margin M.

a) Loads Q b) Strength S c) Safety margin $M = S - Q$

In such a frequency curve (Fig. 1-14a), the area under the curve between two abscissas, such as loads Q_1 and Q_2, represents the probability of occurrence of loads Q of magnitude $Q_1 < Q < Q_2$. A specified service load Q_d for design is selected conservatively in the upper region of Q in the distribution curve, as shown. The probability of occurrence of loads larger than Q_d is then given by the shaded area to the right of Q_d. It is seen that this specified service load is considerably larger than the mean load $\overline{Q}$ acting on the structure. This mean load is much more typical of average load conditions than the design load Q_d.

1.4.2 Strength

The strength of a structure depends on the strength of the materials from which it is made. For this purpose, minimum material strengths are specified in standardized ways. Actual material strengths cannot be known precisely and therefore also constitute random variables. <u>Structural strength depends, furthermore, on the care with which a structure is built, which in turn reflects the quality of supervision and inspection.</u>[7] Member sizes may differ from specified dimensions, reinforcement may be out of position, poorly placed concrete may show voids, etc.

Strength of the entire structure or of a population of repetitive structures, e. g., highway overpasses, can also be considered a random variable with a probability density function of the type shown in Fig. 1-14b. As in the case of loads, the exact form of this function cannot be known but can be approximated from known data, such as statistics of actual, measured materials and member strengths and similar information. Considerable information of this type has been, or is being, developed and used.

1.4.3 Structural Safety

A given structure has a *safety margin* M if

$$M = S - Q > 0 \tag{1-1}$$

i. e., if the strength of the structure is larger than the load acting on it. Since S and Q are random variables, the safety margin $M = S - Q$ is also a random variable. A plot of the probability function of M may appear as in Fig. 1-14c. Failure occurs when M is less than zero. Thus, the probability of failure is represented by the shaded area in the figure.

Even though the precise form of the probability density functions for S and Q, and therefore for M, is not known, much can be achieved in the way of a rational approach to structural safety. One such approach is to require that the mean safety margin M be a specified number β of standard deviations σ_m above zero. It can be demonstrated that this results in the requirement that

$$\psi_s \overline{S} \geq \psi_L \overline{Q} \tag{1-2}$$

where ψ_s is a partial safety coefficient smaller than one applied to the mean strength $\overline{S}$ and ψ_L is a partial safety coefficient larger than one applied to the mean load $\overline{Q}$. The magnitude of each partial safety coefficient depends on the variance of the quantity to which it applies, S or Q, and on the chosen value of β, the reliability index of the structure. As a general guide, a value of the safety index β between 3 and 4 corresponds to a probability of failure of the order of 1 : 100000. The value of β is often established by calibration against well-proved and established designs.

In practice, it is more convenient to introduce partial safety coefficients with respect to code-specified loads which, as already noted, considerably exceed average values, rather than with respect to mean loads as in Eq. (1-2); similarly, the partial safety coefficient for strength is applied to nominal strength generally computed somewhat conservatively, rather than to mean strengths as in

Eq. (1-2). A restatement of the safety requirement in these terms is

$$\phi S_n \geq \gamma Q_d \tag{1-3a}$$

in which ϕ is a strength reduction factor applied to nominal strength S_n and γ is a load factor applied to calculated or code-specified design loads Q_d. Furthermore, recognizing the differences in variability between, say, dead loads D and live loads L, it is both reasonable and easy to introduce different load factors for different types of loads. The preceding equation can thus be written

$$\phi S_n \geq \gamma_d D + \gamma_l L \tag{1-3b}$$

in which γ_d is a load factor somewhat greater than 1.0 applied to the calculated dead load D and γ_l is a larger load factor applied to the code-specified live load L. When additional loads, such as the wind load W, are to be considered, the reduced probability that maximum dead, live, and wind or other loads will act simultaneously can be incorporated by using modified load factors such that

$$\phi S_n \geq \alpha(\gamma_d D + \gamma_l L + \gamma_w W + \cdots) \tag{1-3c}$$

Present U. S. design specifications follow the format of Eq. (1-3b) and Eq. (1-3c).

1.5 Design Basis

The single most important characteristic of any structural member is its actual strength, which must be large enough to resist, with some margin to spare, all foreseeable loads that may act on it during the life of the structure, without failure or other distress.[8] It is logical, therefore, to proportion members, i.e., to select concrete dimensions and reinforcement, so that member strengths are adequate to resist forces resulting from certain hypothetical overload stages, significantly above loads expected actually to occur in service.[9] This design concept is known as strength design.

For reinforced concrete structures at loads close to and at failure, one or both of the materials, concrete and steel, are invariably in their nonlinear inelastic range. That is, concrete in a structural member reaches its maximum strength and subsequent fracture at stresses and strains far beyond the initial elastic range in which stresses and strains are fairly proportional. Similarly, steel close to and at failure of the member is usually stressed beyond its elastic domain into and even beyond the yield region. Consequently, the nominal strength of a member must be calculated on the basis of this inelastic behavior of the materials.

A member designed by the strength method must also perform in a satisfactory way under normal service loading. For example, beam deflections must be limited to acceptable values, and the number and width of flexural cracks at service loads must be controlled. Serviceability limit conditions are an important part of the total design, although attention is focused initially on strength.

Historically, members were proportioned so that stresses in the steel and concrete resulting from normal service loads were within specified limits. These limits, known as allowable stresses, were only fractions of the failure stresses of the materials. For members proportioned on such a service load basis, the margin of safety was provided by stipulating allowable stresses under service loads

Unit 1 Introduction to Reinforced Concrete Design

that were appropriately small fractions of the compressive concrete strength and the steel yield stress.[10] We now refer to this basis for design as service load design. Allowable stresses, in practice, were set at about one-half the concrete compressive strength and one-half the yield stress of the steel.

Because of the difference in realism and reliability, the strength design method has displaced the older service load design method. However, the older method provides the basis for some serviceability checks and is the design basis for many older structures. Throughout this text, strength design is presented almost exclusively.

 New Words and Expressions

  plasticizer *n.* 塑化剂（砂浆），增强剂（塑料），柔韧剂
  air-entraining agent *n.* 加气剂
  strand *n.* 钢筋（丝）束，钢绞线
  wire *n.* 线材，钢丝索
  monolithic *adj.* 整体的
  joist *n.* 搁栅，小梁，托梁
  flared column top 喇叭形（漏斗式）柱顶
  esthetic *adj.* 同 aesthetic，美学的，审美的，艺术的
  hyperbolic paraboloid 双曲抛物面
  spnitary dome 球形屋顶，球形穹顶
  sanitary *adj.* 关于环境卫生的，清洁的
  suspended load 悬挂荷载
  stair tread 楼梯踏步板
  serviceability *n.* 适用性
  abscissas *n.* 横坐标
  overload *n.* 超载，超负荷

1. Additional water, over and above that needed for this chemical reaction, is necessary to give the mixture the workability that enables it to fill the forms and surround the embedded reinforcing steel prior to hardening.

本句难点解析：句子主体是 Additional water is necessary to give the mixture the workability，即"多余的水分是为了保证混凝土的和易性"。关键词是 Additional "多余的"，workability "和易性"。其中 over and above that needed for this chemical reaction 是用来修饰 additional water 的，表达"多余的水分"中的"多余"所指的方面。that enables it to fill the forms and surround the embedded reinforcing steel prior to hardening 这句话为 workability 的定语从句，表示"保证和易性是为了让混合物（即混凝土）在硬化前能成型包裹住钢筋"。

本句大意如下：超过化学作用所需的额外的水用来提供混合物的和易性，使之在硬化之

前能够填入模板以及包裹嵌入的钢筋。

2. These properties depend to a very substantial degree on the proportions of the mix, on the thoroughness with which the various constituents are intermixed, and on the conditions of humidity and temperature in which the mix is maintained from the moment it is placed in the forms until it is fully hardened.

本句难点解析：句子主体是 These properties depend on the proportions, the thoroughness and the conditions of humidity and temperature，即"这些性质取决于混合的彻底性与温湿度的条件"。其中 to a very substantial degree 作为状语修饰 depend，表示"很大程度上"。with which the various constituents are intermixed 以及 in which the mix is maintained from the moment it is placed in the forms until it is fully hardened 则分别作为定语从句修饰 thoroughness 和 conditions，表示"彻底性"所指代的是几种材料的混合情况，"条件"所指代的是混合物从放置进模具直至成型的养护条件。

本句大意如下：这些特性很大程度上依赖于混合比例，不同组成成分搅拌的充分性，以及混合物在浇注后直到硬化前的湿度和温度。

3. The facility with which, while plastic, it can be deposited and made to fill forms or molds of almost any practical shape is one of these factors.

本句难点解析：句子主体是 The facility is one of these factors，即"便捷性是这些因素的其中之一"。with which it can be deposited and made to fill forms or molds of almost any practical shape 作为定语从句修饰 facility，表示"便捷性"所指代的是混凝土能放置在几乎任何实际模具中成型。而 while plastic 则是作为状语修饰 deposit 以及 make，表示放置的时候混凝土应处于塑性状态。

本句大意如下：当混凝土处于塑性状态时，它可以很容易地浇注并填充到任何形式的模板或磨具中这一优点便是其中一个因素。

4. The steel, in the form of wires, strands, or bars, is embedded in the concrete under high tension that is held in equilibrium by compressive stresses in the concrete after hardening.

本句难点解析：句子主体是 The steel is embedded in the concrete under high tension。其中 in the form of wires, strands, or bars 作为修饰成分说明钢筋的形式，that is held in equilibrium by compressive stresses in the concrete after hardening 作为定语从句修饰拉应力，表明与钢筋拉应力相平衡的是混凝土中的压应力。

本句大意如下：钢筋的形式通常是钢线、钢绞线或是钢筋，其在较高的拉力作用下埋入混凝土中，且该拉力与硬化后混凝土的受压应力保持平衡。

5. If the strength of a structure, built as designed, could be predicted accurately, and if the loads and their internal effects (moments, shears, axial forces) were known accurately, safety could be ensured by providing a carrying capacity just barely in excess of the known loads.

本句大意如下：如果按设计建造的结构，其结构强度，荷载以及对应的内力（弯矩，剪力，轴力）都能被准确预测的话，则仅需使其承载力比已知荷载稍大便可以确保结构安全。

6. The exact form of this distribution curve, for any particular type of loading such as office loads, can be determined only on the basis of statistical data obtained from large-scale load surveys.

Unit 1　Introduction to Reinforced Concrete Design

本句难点解析：句子主体是 The exact form can be determined on the basis of statistical data。其中 for any particular type of loading such as office loads 修饰 curve，obtained from large-scale load surveys 作为定语从句修饰 data。

本句大意如下：对于任意一种具体的荷载，比如办公室荷载，其分布曲线的确定仅能以大规模荷载调查所得的统计数据作为基础。

7. Structural strength depends, furthermore, on the care with which a structure is built, which in turn reflects the quality of supervision and inspection.

本句难点解析：句子主体是 Structural strength depends on the care。with which a structure is built 作为定语从句修饰 care，which in turn reflects the quality of supervision and inspection 同样作为定语从句修饰 care。

本句大意如下：此外，结构的强度还取决于建造时的认真程度，这反过来也反映了监管及检查的质量。

8. The single most important characteristic of any structural member is its actual strength, which must be large enough to resist, with some margin to spare, all foreseeable loads that may act on it during the life of the structure, without failure or other distress.

本句难点解析：which 后面的从句作为定语从句修饰 strength，从句主体是 which must be large enough to resist all foreseeable loads，其中 with some margin to spare 作为伴随状语从句修饰前句，表示"强度应该有所富余"，that may act on it during the life of the structure 作为定语从句修饰 loads，without failure or other distress 作为伴随状语从句修饰主句。

本句大意如下：结构构件最重要的特性是其真实强度：强度必须足够大且有所富余，能足以抵抗构件工作过程中可能出现的所有荷载，且不发生破坏。

9. It is logical, therefore, to proportion members, i. e., to select concrete dimensions and reinforcement, so that member strengths are adequate to resist forces resulting from certain hypothetical overload stages, significantly above loads expected actually to occur in service.

本句难点解析：句子主体是 It is logical to proportion members。to select concrete dimensions and reinforcement 这一句可根据前面的 i. e. 得知其是解释 proportion members 的含义。

本句大意如下：应该合理地选择构件尺寸，即混凝土的截面尺寸以及钢筋尺寸，才能让构件有足够的强度抵抗可能出现的超载情况，即大幅超出正常使用期间预期荷载的情况。

10. For members proportioned on such a service load basis, the margin of safety was provided by stipulating allowable stresses under service loads that were appropriately small fractions of the compressive concrete strength and the steel yield stress.

本句难点解析：句子中 on such a service load basis 作为状语修饰 proportioned，表明"构件尺寸按正常使用荷载进行定义"。

本句大意如下：对于以正常使用荷载来定义截面尺寸的构件来说，安全度通过规定正常使用荷载下的应力来实现，限制该应力仅为混凝土抗压强度及钢筋屈服应力的一小部分。

Translate the following phrases in to Chinese.

1. stonelike material
2. reinforcing steel
3. Construction site
4. heavyweight aggregate
5. constituent material
6. flexural member
7. prestressed concrete
8. base shear
9. design configuration
10. average value

Translate the following Sentences in to Chinese.

1. Concrete is a stonelike material obtained by permitting a carefully proportioned mixture of cement, sand and gravel or other aggregates, and water to harden in forms of the shape and dimensions of the desired structure.

2. To offset this limitation, it was found possible, in the second half of the nineteenth century, to use steel with its high tensile strength to reinforce concrete, chiefly in those places where its low tensile strength would limit the carrying capacity of the member.

3. The resulting combination of two materials, known as reinforced concrete, combines many of the advantages of each: the relatively low cost, good weather and fire resistance, good compressive strength, and excellent formability of concrete and the high tensile strength and much greater ductility and toughness of steel.

4. Dead loads are those that are constant in magnitude and fixed in location throughout the lifetime of the structure.

5. Seismic forces may be found for a particular structure by elastic or inelastic dynamic analysis, considering expected ground accelerations and the mass, stiffness, and damping characteristics of the construction.

Unit 2

Introduction to Prestressed Concrete

2.1 Introduction

预应力章节逻辑框架

Modern structural engineering tends to progress toward more economic structures through gradually improved methods of design and the use of higher strength materials. This results in a reduction of cross-sectional dimensions and consequent weight savings. Such developments are particularly important in the field of reinforced concrete, where the dead load represents a substantial part of the total design load. Also, in multistory buildings, any saving in depth of members, multiplied by the number of stories, can represent a substantial saving in total height, load on foundations, length of heating and electrical ducts, plumbing risers, and wall and partition surfaces.[1]

Significant savings can be achieved by the use of high-strength concrete and steel in conjunction with present-day design methods, which permit an accurate appraisal of member strength. However, there are limitations to this development, due mainly to the interrelated problems of cracking and deflection at service loads. The efficient use of high-strength steel is limited by the fact that the amount of cracking (width and number of cracks) is proportional to the strain, and therefore the stress, in the steel. Although a moderate amount of cracking is normally not objectionable in structural concrete, excessive cracking is undesirable in that it exposes the reinforcement to corrosion, it may be visually offensive, and it may trigger a premature failure by diagonal tension. The use of high-strength materials is further limited by deflection considerations, particularly when refined analysis is used. The slender members that result may permit deflections that are functionally or visually unacceptable. This is further aggravated by cracking, which reduces the flexural stiffness of members.

These limiting features of ordinary reinforced concrete have been largely overcome by the development of prestressed concrete. A prestressed concrete member can be defined as one in which there have been introduced internal stresses of such magnitude and distribution that the stresses resulting from the given external loading are counteracted to a desired degree. Concrete is basically a compressive material, with its strength in tension being a low and unreliable value. Prestressing applies a precompression to the member that reduces or eliminates undesirable tensile stresses that would otherwise be present. Cracking under service loads can be minimized or even avoided entirely. Deflections may be limited to an acceptable value; in fact, members can be designed to have zero deflection under the combined effects of service load and prestress force. Deflection and crack control, achieved through prestressing, permit the engineer to make use of efficient and economical high-strength steels in the form of strands, wires, or bars, in conjunction with concretes of much higher strength than normal.[2] Thus, prestressing results in overall improvement in performance of structural concrete used for ordinary loads and spans and extends the range of application far beyond old limits, leading not only to much longer spans than previously thought possible, but also permitting innovative new structural forms to be employed.[3]

2.2　Effects of Prestressing

There are at least three alternative ways to look at the prestressing of concrete: (a) as a method of achieving concrete stress control, by which the concrete is precompressed so that tension normally resulting from the applied loads is reduced or eliminated; (b) as a means for introducing *equivalent loads* on the concrete member so that the effects of the applied loads are counteracted to the desired degree; and (c) as a special variation of reinforced concrete in which prestrained high-strength steel is used, usually in conjunction with high-strength concrete. Each of these viewpoints is useful in the analysis and design of prestressed concrete structures, and they will be illustrated in the following paragraphs.

2.2.1　Concrete Stress Control by Prestressing

Many important features of prestressed concrete can be demonstrated by simple examples. Consider first the plain, unreinforced concrete beam shown in Fig. 2-1a. It carries a single concentrated load at the center of its span (The self-weight of the member will be neglected here). As the load W is gradually applied, longitudinal flexural stresses are induced. If the concrete is stressed only within its elastic range, the flexural stress distribution at midspan will be linear, as shown.

At a relatively low load, the tensile stress in the concrete at the bottom of the beam will reach the tensile strength of the concrete f_r, and a crack will form. Because no restraint is provided against upward extension of the crack, the beam will collapse without further increase of load.

Now consider an otherwise identical beam, shown in Fig. 2-1b, in which a longitudinal axial force P is introduced prior to the vertical loading. The longitudinal prestressing force will produce a uniform axial compression $f_c = P/A_c$, where A_c is the cross-sectional area of the concrete. The force can be adjusted in magnitude so that when the transverse load Q is applied, the superposition of stresses due to P and Q will result in zero tensile stress at the bottom of the beam as shown. Tensile stress in the concrete may be eliminated in this way or reduced to a specified amount.

But it would be more logical to apply the prestressing force near the bottom of the beam, to compensate more effectively for the load-induced tension.[4] A possible design specification, for example, might be to introduce the maximum compression at the bottom of the beam without causing tension at the top, when only the prestressing force acts.[5] It is easily shown that, for a beam with a rectangular cross section, the point of application of the prestressing force should be at the lower third point of the section depth to achieve this. The force P, with the same value as before, but applied with eccentricity $e = h/6$ relative to the concrete centroid, will produce a longitudinal compressive stress distribution varying linearly from zero at the top surface to a maximum of $2f_c = P/A_c + Pec_2/I_c$ at the bottom, where f_c is the concrete stress at the concrete centroid, c_2 is the distance from the concrete centroid to the bottom of the beam, and I_c is the moment of inertia of the cross section. This is shown in Fig. 2-1c. The stress at the bottom will be exactly twice the value produced before by axial prestressing.

Consequently, the transverse load can now be twice as great as before, or $2Q$, and still cause no tensile stress. In fact, the final stress distribution resulting from the superposition of load and prestressing force in Fig. 2-1c is identical to that of Fig. 2-1b, with the same prestressing force, although the load is twice as great. The advantage of eccentric prestressing is obvious.

The methods by which concrete members are prestressed will be discussed in Section 2.3. For present purposes, it is sufficient to know that one practical method of prestressing uses high-strength steel tendons passing through a conduit embedded in the concrete beam. The tendon is anchored, under high tension, at both ends of the beam, thereby causing a longitudinal compressive stress in the concrete. The prestress force of Fig. 2-1b and c could easily have been applied in this way.

A significant improvement can be made, however, by using a prestressing tendon with variable eccentricity with respect to the concrete centroid, as shown in Fig. 2-1d. The load $2Q$ produces a bending moment that varies linearly along the span, from zero at the supports to maximum at midspan. Intuitively, one suspects that the best arrangement of prestressing would produce a countermoment that acts in the opposite sense to the load-induced moment and that would vary in the same way. This would be achieved by giving the tendon with an eccentricity that varies linearly, from zero at the supports to maximum at midspan. This is shown in Fig. 2-1d. The stresses at midspan are the same as those in Fig. 2-1c, both when the load $2Q$ acts and when it does not. At the supports, where only the prestress force with zero eccentricity acts, a uniform compression stress f_c is obtained as shown.

For each characteristic load distribution, there is a best tendon profile that produces a prestress moment diagram that corresponds to that of the applied load. If the prestress countermoment is made exactly equal and opposite to the load-induced moment, the result is a beam that is subject only to uniform axial compressive stress in the concrete all along the span. Such a beam would be free of flexural cracking, and theoretically it would not be deflected up or down when that particular load is in place, compared to its position as originally cast. Such a result would be obtained for a load of $(1/2) \times 2Q = Q$, as shown in Fig. 2-1e, for example.

Some important conclusions can be drawn from these simple examples as follows:

1. Prestressing can control or even eliminate concrete tensile stress for specified loads.

2. Eccentric prestress is usually much more efficient than concentric prestress.

3. Variable eccentricity is usually preferable to constant eccentricity, from the viewpoints of both stress control and deflection control.

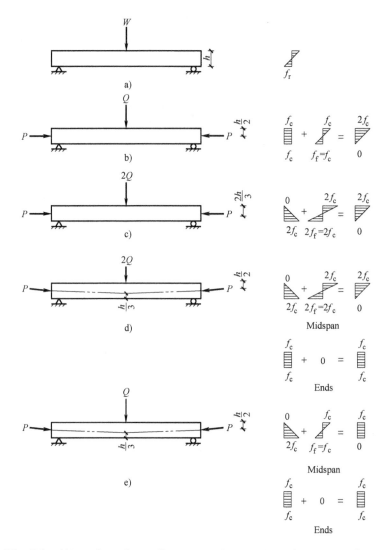

Fig. 2-1 Alternative schemes for prestressing a rectangular concrete beam.
a) Plain concrete beam b) Axially prestressed beam c) Eccentrically prestressed beam
d) Beam with variable eccentricity e) Balanced load stage for beam with variable eccentricity

2.2.2 Equivalent Loads

预应力
等效荷载 1

The effect of a change in the vertical alignment of a prestressing tendon is to produce a vertical force on the concrete beam. That force, together with the prestressing force acting at the ends of the beam through the tendon anchorages, can be looked upon as a system of external loads.[6]

In Fig. 2-2a, for example, a tendon that applies force P at the centroid of the concrete section at the ends of a beam and that has a uniform slope at angle θ between the ends and midspan introduces a transverse force $2P \sin \theta$ at the point of change of slope at midspan. At the anchorages, the vertical component of the prestressing force is $P\sin\theta$ and the horizontal component is $P\cos \theta$. The horizontal component is very nearly equal to P for the usual flat slope angles. The moment diagram for the beam of Fig. 2-2a is seen to have the same form as that for any center-loaded simple span.

The beam of Fig. 2-2b, with a curved tendon, is subject to a vertical upward load from the tendon as well as the forces P at each end. The exact distribution of the load depends on the profile of the tendon. A tendon with a parabolic profile, for example, will produce a uniformly distributed load. In this case, the moment diagram will be parabolic, as it is for a uniformly loaded simple span.

If a straight tendon is used with constant eccentricity, as shown in Fig. 2-2c, there are no vertical forces on the concrete, but the beam is subject to a moment Pe at each end, as well as the axial force P, and a diagram of constant moment results.

The end moment must also be accounted for in the beam of Fig. 2-2d, in which a parabolic tendon is used that does not pass through the concrete centroid at the ends of the span. In this case, a uniformly distributed upward load plus end anchorage forces are produced, as shown in Fig. 2-2b, but in addition, the end moments $M = Pe\cos\theta$ must be accounted for.

预应力
等效荷载 2

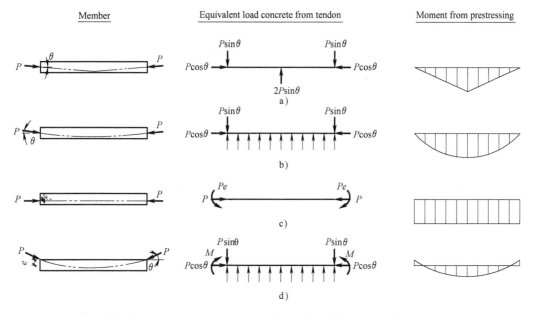

Fig. 2-2 Equivalent loads and moments produced by prestressing tendons.

It may be evident that for any arrangement of applied loads, a tendon profile can be selected so that the equivalent loads acting on the beam from the tendon are just equal and opposite to the applied loads. The result would be a state of pure compressive stress in the concrete, as discussed in somewhat different terms in reference to stress control and Fig. 2-1e. An advantage of the equivalent load concept is that it leads the designer to select what is probably the best tendon profile for a particular loading.

2.2.3 Prestressed Concrete as a Variation of Reinforced Concrete

In the descriptions of the effects of prestressing in the paragraphs above, it was implied that the prestress force remained constant as the vertical load was introduced, that the concrete responded elastically, and that no concrete cracking occurred.[7] These conditions may prevail up to about the service load level, but if the loads should be increased much beyond that, flexural tensile stresses will eventually exceed the modulus of rupture and cracks will form. Loads can usually be increased much beyond the cracking load in well-designed prestressed beams.

Eventually both the steel and concrete at the cracked section will be stressed into the inelastic range. The condition at incipient failure is shown in Fig. 2-3, which shows a beam carrying a factored load equal to some multiple of the expected service load. The beam undoubtedly would be in a partially cracked state; a possible pattern of flexural cracking is shown in Fig. 2-3a.

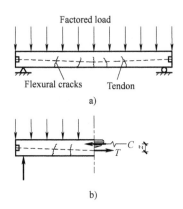

Fig. 2-3 Prestressed concrete beam at load near flexural failure.
a) Beam with factored load applied
b) Equilibrium of forces on left half of beam

At the maximum moment section, only the concrete in compression is effective, and all of the tension is taken by the steel. The external moment from the applied loads is resisted by the internal forces couple $Cz = Tz$. The behavior at this stage is almost identical to that of an ordinary reinforced concrete beam at overload. The main difference is that the very high strength steel used must be prestrained before loads are applied to the beam; otherwise, the high steel stresses would produce excessive concrete cracking and large beam deflections.

Each of the three viewpoints described—concrete stress control, equivalent loads, and reinforced concrete using prestrained steel—is useful in the analysis and design of prestressed concrete beams, and none of the three is sufficient in itself. Neither an elastic stress analysis nor an equivalent load analysis provides information about strength or safety margin. However, the stress analysis is helpful in predicting the extent of cracking, and the equivalent load analysis is often the best way to calculate deflections. Strength analysis is essential to evaluate safety against collapse, but it tells nothing about cracking or deflections of the beam under service conditions.

2.3 Sources of Prestress Force

Prestress can be applied to a concrete member in many ways. Perhaps the most obvious method of precompressing is the use of jacks reacting against abutments, as shown in Fig. 2-4a. Such a scheme has been employed for large projects. Many variations are possible, including replacing the jacks with compression struts after the desired stress in the concrete is obtained or using inexpensive jacks, that remain in place in the structure, in some cases with a cement grout used as the hydraulic fluid.[8] The principal difficulty associated with such a system is that even a slight movement of the abutments will drastically reduce the prestress force.

In most cases, the same result is more conveniently obtained by tying the jack bases together with wires or cables, as shown in Fig. 2-4b. These wires or cables may be external, located on each side of the beam; more usually they are passed through a hollow conduit embedded in the concrete beam. Usually, one end of the prestressing tendon is anchored, and all the force is applied at the other end. After attainment of the desired prestress force, the tendon is wedged against the concrete and the jacking equipment is removed for reuse. Note that in this type of prestressing, the entire system is self-contained and is independent of relative displacement of the supports.

Another method of prestressing that is widely used is illustrated by Fig. 2-4c. The prestressing strands are tensioned between massive abutments in a casting yard prior to placing the concrete in the beam forms. The concrete is placed around the tensioned strands, and after the concrete has attained sufficient strength, the jacking pressure is released. This transfers the prestressing force to the concrete by bond and friction along the strands, chiefly at the outer ends.

Other means for introducing the desired prestressing force have been attempted on an experimental basis. Thermal prestressing can be achieved by preheating the steel by electrical or other means. Anchored against the ends of the concrete beam while in the extended state, the steel cools and tends to contract. The prestress force is developed through the restrained contraction. The use of expanding cement in concrete members has been tried with varying success. The volumetric expansion, restrained by steel strands or by fixed abutments, produces the prestress force.

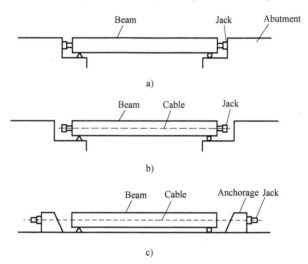

Fig. 2-4 Prestressing methods.
a) Post-tensioning by jacking against abutments   b) Post-tensioning with jacks reacting against beam
c) Pretensioning with tendon stressed between fixed external anchorages

Most of the patented systems for applying prestress in current use are variations of those shown in Fig. 2-4b and c. Such systems can generally be classified as pretensioning or post-tensioning systems. In the case of pretensioning, the tendons are stressed before the concrete is placed, as in Fig. 2-4c. This system is well suited for mass production, since casting beds can be made several hundred feet long, the entire length cast at once, and individual beams cut to the desired length in a single casting. Fig. 2-5 shows workers using a hydraulic jack to tension strands at the anchorage of a long pretensioning bed. Although each tendon is individually stressed in this case, large capacity jacks are often used to tension all strands simultaneously.

In post-tensioned construction, shown in Fig. 2-4b, the tendons are tensioned after the concrete is placed and has acquired its strength. Usually, a hollow conduit or sleeve is provided in the beam, through which the tendon is passed. In some cases, hollow box-section beams are used. The jacking force is usually applied against the ends of the hardened concrete, eliminating the need for massive abutments. In Fig. 2-6, six tendons, each consisting of many individual strands, are being post-tensioned sequentially using a portable hydraulic jack.

Fig. 2-5 Massive strand jacking abutment at the end of a long pretensioning bed (Courtesy of Concrete Technology Corporation).

Fig. 2-6 Post-tensioning a bridge girder: use a portable jack to stress multistrand tendons (Courtesy of Concrete Technology Corporation).

A large number of particular systems, steel elements, jacks, and anchorage fittings have been developed in this country and abroad, many of which differ from each other only in minor details. As far as the designer of prestressed concrete structures is concerned, it is unnecessary and perhaps even undesirable to specify in detail the technique that is to be followed and the equipment to be used. It is frequently best to specify only the magnitude and line of action of the prestress force. The contractor is then free, in bidding the work, to receive quotations from several different prestressing subcontractors, with resultant cost savings. It is evident, however, that the designer must have some knowledge of the details of the various systems contemplated for use, so that in selecting cross-sectional dimensions, any one of several systems can be accommodated.[9]

2.4 Prestressing Steels

Early attempts at prestressing concrete were unsuccessful because steel of ordinary structural strength was used. The low prestress obtainable in such rods was quickly lost due to shrinkage and creep in the concrete.

Such changes in length of concrete have much less effect on prestress force if that force is obtained using highly stressed steel wires or cables. In Fig. 2-7a, a concrete member of length L is prestressed using steel bars of ordinary strength stressed to 24000psi (1psi = 6.89kPa). With $E_s = 29 \times 10^6$ psi, the unit strain ε_s, required to produce the desired stress in the steel of 24000psi is

$$\varepsilon_s = \frac{\Delta L}{L} = \frac{f_s}{E_s} = \frac{24000\text{psi}}{29 \times 10^6 \text{psi}} = 8.0 \times 10^{-4}.$$

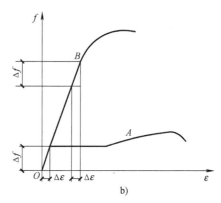

However, the long-term strain in the concrete due to shrinkage and creep alone, if the prestress forces were maintained over a long period, would be on the order of 8.0×10^{-4} and would be sufficient to completely relieve the steel of all stress.

Fig. 2-7 Loss of prestress due to concrete shrinkage and creep.

Alternatively, suppose that the beam is prestressed using high tensile steel stressed to 150000psi. The elastic modulus of steel does not vary greatly, and the same value of 29×10^{-6} psi will be assumed here. Then in this case, the unit strain required to produce the desired stress in the steel is

预应力钢

$$\varepsilon_s = \frac{150000\text{psi}}{29 \times 10^6 \text{psi}} = 51.7 \times 10^{-4}$$

If shrinkage and creep strain are the same as before, the net strain in the steel after these losses is

$$\varepsilon_{s,\text{net}} = (51.7 - 8.0) \times 10^{-4} = 43.7 \times 10^{-4}$$

and the corresponding stress after losses is

$$f_s = \varepsilon_{s,\text{net}} E_s = 4.37 \times 10^{-4} \times 29 \times 10^6 \text{psi} = 127000\text{psi}$$

This represents a stress loss of about 15 percent, compared with 100 percent loss in the beam using ordinary steel. It is apparent that the amount of stress lost because of shrinkage and creep is dependent upon the original stress in the steel. Therefore, the higher the original stress, the lower the percentage loss. This is illustrated graphically by the stress-strain curves of Fig. 2-7b. Curve A is representative of ordinary reinforcing bars, with a yield stress of 60000psi, while curve B represents high tensile steel, with an ultimate stress of 250000psi. The stress change Δf resulting from a certain change in strain $\Delta \varepsilon$ is seen to have much less effect when high steel stress levels are attained. Prestressing of concrete is therefore practical only when steels of very high strength are used.

Prestressing steel is most commonly used in the form of individual wires, stranded cable made up of seven wires, and alloy-steel bars.

The tensile stress permitted by ACI Code 18.5 in prestressing wires, strands, or bars is dependent upon the stage of loading. When the jacking force is first applied, a stress of $0.80 f_{pu}$ or $0.94 f_{py}$ is allowed, whichever is smaller, where f_{pu} is the ultimate strength of the steel and f_{py} is the yield strength. Immediately after transfer of prestress force to the concrete, the permissible stress is $0.74 f_{pu}$ or $0.82 f_{py}$, whichever is smaller (except at post-tensioning anchorages where the stress is limited to $0.70 f_{pu}$). The justification for a higher allowable stress during the stretching operation is that the steel stress is known quite precisely at this stage. Hydraulic jacking pressure and total steel strain are quantities that are easily measured. In addition, if an accidentally deficient tendon should break, it can be replaced; in effect, the tensioning operation is a performance test of the material. The lower values of allowable stress apply after elastic shortening of the concrete, frictional loss, and anchorage slip have taken place, when service loads may be applied. The steel stress is further reduced during the life of the member due to shrinkage and creep in the concrete and relaxation in the steel.

The strength and other characteristics of prestressing wire, strands, and bars vary somewhat between manufacturers, as do methods of grouping tendons and anchoring them.

■ 2.5 Concrete for Prestressed Construction

Ordinarily, concrete of substantially higher compressive strength is used for prestressed structures than for those constructed of ordinary reinforced concrete. Most prestressed construction in the United States at present is designed for a compressive strength between 5000psi and 6000psi. There are several reasons for this:

1. High-strength concrete normally has a higher modulus of elasticity (see Fig. 2-3). This means a reduction in initial elastic strain under application of prestress force and a reduction in creep strain, which is approximately proportional to elastic strain. This results in a reduction in loss of prestress.

2. In post-tensioned construction, high bearing stresses result at the ends of beams where prestressing force is transferred from the tendons to anchorage fittings, which bear directly against the concrete. This problem can be met by increasing the size of the anchorage fitting or by increasing the bearing capacity of the concrete, by increasing its compressive strength. The latter is usually more economical.

3. In pretensioned construction, where transfer by bond is customary, the use of high-strength concrete will permit the development of higher bond stresses.

4. A substantial part of the prestressed construction in the United States is precast, with the concrete mixed, placed, and cured under carefully controlled conditions that facilitate obtaining higher strengths.

The strain characteristics of concrete under short-time and sustained loads assume an even greater importance in prestressed structures than in reinforced concrete structures because of the influence of strain on loss of prestress force.[10] Strains due to stress, together with volume changes due to shrinkage and temperature changes, may have considerable influence on prestressed structures.

Table 2-1 Permissible stresses in concrete in prestressed flexural members

Concrete in prestressed flexural members	Permissible stresses
1. Stresses in concrete immediately after prestress transfer, before time-dependent prestress losses, shall not exceed the following:	
a. Extreme fiber stress in compression	$0.60 f'_{ci}$
b. Extreme fiber stress in tension except as permitted in (c)	$3\sqrt{f'_{ci}}$
c. Extreme fiber stress in tension at ends of simply supported members	$6\sqrt{f'_{ci}}$
Where computed tensile stresses exceed these values, bonded auxiliary reinforcement (nonprestressed or prestressed) shall be provided in the tensile zone to resist the total tensile force in the concrete computed with the assumption of an uncracked section	
2. Stress in concrete at service loads, after allowance for all prestress losses, shall not exceed the following:	
a. Extreme fiber stress in compression	$0.45 f'_c$
b. Extreme fiber stress in tension in precompressed tensile zone	$6\sqrt{f'_c}$
c. Extreme fiber stress in tension in precompressed tensile zone of members, except two-way slab systems, where analysis based on transformed cracked sections and on bilinear moment-deflection relationships shows that immediate and long-time deflections comply with restrictions stated elsewhere in the ACI Code	$12\sqrt{f'_c}$
3. Permissible stresses in concrete given above may be exceeded if it is shown by test or analysis that performance will be impaired	

As for prestressing steels, the allowable stresses in the concrete, according to ACI Code 18.4, depend upon the stage of loading. These stresses are given in Table 2-1. Here f'_{ci} is the compressive strength of the concrete at the time of initial pressure, and the f'_c the specified compressive strength of the concrete.

New Words and Expressions

plumbing　　n.　　管道工程，卫生工程

riser　　n.　　竖管，井管，溢水管，提升井，提升装置

tendon　　n.　　钢筋，钢筋束

conduit　　n.　　管道，导线管

profile   *n.*   立面图，剖面图，外形，模型
incipient   *adj.*   早期的，刚出现的
factored load   极限设计荷载
jack   *n.*   千斤顶
abutment   *n.*   支座，支墩，拱座
strut   *n.*   支撑，压杆，对角撑
grout   *n.*   浆，水泥浆；*V.* 灌浆
sleeve   *n.*   套筒（管），管接头
bid   *n.*   投标，标书
quotation   *n.*   报价单，估计单，行情，应用
contemplate   *V.*   仔细考虑，沉思，期待

Notes

1. Also, in multistory buildings, any saving in depth of members, multiplied by the number of stories, can represent a substantial saving in total height, load on foundations, length of heating and electrical ducts, plumbing risers, and wall and partition surfaces.

本句难点解析：句子中 multiplied by the number of stories 作为定语修饰 saving，句子的主谓宾分别是 saving，represent，saving，主语 saving 包括 depth of members，宾语则为 total height，load on foundations，length of heating and electrical ducts，plumbing risers，and wall and partition surfaces 并列。

本句大意如下：同样，在多层建筑中，构件高度的减小值乘以总层数，可以代表总高度、基础上的荷载、供热供电管线长度、管道长度、墙以及隔墙面的减小值。

2. Deflection and crack control, achieved through prestressing, permit the engineer to make use of efficient and economical high-strength steels in the form of strands, wires, or bars, in conjunction with concretes of much higher strength than normal.

本句难点解析：句子中 achieved through prestressing 作为定语修饰 deflection and crack control，句子的主体是 Deflection and crack control permit the engineer to make use of steels。

本句大意如下：通过预应力来实现挠度及裂缝控制的技术使得工程师可以采用高效经济的高强钢筋（以钢绞线，钢线或钢筋形式），并可以协同使用高强混凝土。

3. Thus, prestressing results in overall improvement in performance of structural concrete used for ordinary loads and spans and extends the range of application far beyond old limits, leading not only to much longer spans than previously thought possible, but also permitting innovative new structural forms to be employed.

本句难点解析：句子中 leading not only to much longer spans than previously thought possible，but permitting innovative new structural forms to be employed 作为定语修饰 prestressing results。

本句大意如下：预应力的应用，使得普通荷载和跨度下的结构混凝土性能得到整体提高，并扩大了其应用范围，不但可以增加结构的跨度，而且可以创造出新的结构形式。

Unit 2 Introduction to Prestressed Concrete

4. But it would be more logical to apply the prestressing force near the bottom of the beam, to compensate more effectively for the load-induced tension.

本句难点解析：句子中 to compensate more effectively for the load-induced tension 为 to 引导的状语从句，表示目的。

本句大意如下：但是将预应力施加在梁底部会更加合理，这样可以更有效地抵抗荷载所引起的拉力。

5. A possible design specification, for example, might be to introduce the maximum compression at the bottom of the beam without causing tension at the top, when only the prestressing force acts.

本句难点解析：句子中 when only the prestressing force acts 作为状语从句修饰前句，句子主体是 A possible design specification might be to introduce the maximum compression at the bottom of the beam。

本句大意如下：举个例子，一个可能的设计方案是当只有预应力作用时，在梁底部产生最大压力但在顶部不产生拉力。

6. That force, together with the prestressing force acting at the ends of the beam through the tendon anchorages, can be looked upon as a system of external loads.

本句难点解析：句子的主体是 That force can be looked upon as a system of external loads。其中 together with the prestressing force acting at the ends of the beam through the tendon anchorages 作为定语修饰 force。

本句大意如下：与预应力一起通过锚具作用在梁端的力，可以被视为一个外部荷载。

7. In the descriptions of the effects of prestressing in the paragraphs above, it was implied that the prestress force remained constant as the vertical load was introduced, that the concrete responded elastically, and that no concrete cracking occurred.

本句难点解析：句子中三个 that 从句均对应于 it was implied，表示推断的内容。

本句大意如下：在上个段落对预应力作用的描述中，可以推断在竖向荷载作用下预应力仍保持不变，混凝土为弹性反应，并且没有混凝土裂缝产生。

8. Many variations are possible, including replacing the jacks with compression struts after the desired stress in the concrete is obtained or using inexpensive jacks, that remain in place in the structure, in some cases with a cement grout used as the hydraulic fluid.

本句大意如下：有许多不同的做法，包括当混凝土获得预期应力后将千斤顶换成压杆，或者采用不贵的千斤顶，直接将其留在结构中，在某些情况下还可以将水泥浆作为液压体。

9. It is evident, however, that the designer must have some knowledge of the details of the various systems contemplated for use, so that in selecting cross-sectional dimensions, any one of several systems can be accommodated.

本句大意如下：然而，设计人员必须掌握可能使用的几种体系的相关知识，才能在选择截面尺寸时推荐采用相应的体系。

10. The strain characteristics of concrete under short-time and sustained loads assume an even greater importance in prestressed structures than in reinforced concrete structures because of the influence of strain on loss of prestress force.

本句难点解析：句子的主体是 The strain characteristics of concrete assume an even greater importance in prestressed structures than in reinforced concrete structures。其中 under short-time and sustained loads 作为定语修饰 concrete。

本句大意如下：混凝土在短期及长期荷载作用下的应变问题对预应力结构产生的影响要大于钢筋混凝土结构，因为应变损失将引起预应力损失。

Translate the following phrases into Chinese.

1. structural engineering
2. longitudinal axial force
3. equivalent load
4. transverse load
5. tendon profile
6. under service condition
7. portable hydraulic jack
8. cross-sectional dimension
9. diagonal bracing
10. extreme fiber stress

Translate the following sentences into Chinese.

1. Modern structural engineering tends to progress toward more economic structures through gradually improved methods of design and the use of higher strength materials.

2. Because no restraint is provided against upward extension of the crack, the beam will collapse without further increase of load.

3. Strength analysis is essential to evaluate safety against collapse, but it tells nothing about cracking or deflections of the beam under service conditions.

4. Some important conclusions can be drawn from these simple examples as follows:

1) Prestressing can control or even eliminate concrete tensile stress for specified loads.

2) Eccentric prestress is usually much more efficient than concentric prestress.

3) Variable eccentricity is usually preferable to constant eccentricity, from the viewpoints of both stress control and deflection control.

5. It may be evident that for any arrangement of applied loads, a tendon profile can be selected so that the equivalent loads acting on the beam from the tendon are just equal and opposite to the applied loads.

Unit 3

Introduction to Steel Structures

3.1 Structural Design

Structural design may be defined as a mixture of art and science, combining the experienced engineer's intuitive feeling for the behavior of a structure with a sound knowledge of the principles of statics, dynamics, mechanics of materials, and structural analysis, to produce a safe economical structure that will serve its intended purpose.[1]

结构设计

Until about 1850, structural design was largely an art relying on intuition to determine the size and arrangement of the structural elements. Early man-made structures essentially conformed to those which could also be observed in nature, such as beams and arches. As the principles governing the behavior of structures and structural materials have become better understood, design procedures have become more scientific.

Computations involving scientific principles should serve as a guide to decision making and not be followed blindly. The art or intuitive ability of the experienced engineer is utilized to make the decisions, guided by the computational results.

3.2 Principles of Design

Design is a process by which an optimum solution is obtained. In this text the concern is with the design of structures-in particular, *steel* structures. In any design, certain criteria must be established to evaluate whether or not an optimum has been achieved. For a structure, typical criteria may be (a) minimum cost; (b) minimum weight; (c) minimum construction time; (d) minimum labor; (e) minimum cost of manufacture of owner's products; and (f) maximum efficiency of operation to owner. Usually several criteria are involved, each of which may require weighting. Observing the above possible criteria, it may be apparent that setting clearly measurable criteria (such as weight and cost) for establishing an optimum frequently will be difficult, and perhaps impossible. In most practical situations, the evaluation must be qualitative.

If a specific objective criterion can be expressed mathematically, then optimization techniques may be employed to obtain a maximum or minimum for the objective function. Optimization procedures and techniques comprise an entire subject that is outside the scope of this text. The criterion of minimum weight is emphasized throughout, under the general assumption that minimum material represents minimum cost. Other subjective criteria must be kept in mind, even though the integration of behavioral principles with design of structural steel elements in this text utilizes only simple objective criteria, such as weight or cost.

The design procedure may be considered to be composed of two parts—functional design and structural framework design. Functional design ensures that intended results are achieved, such as (a) adequate working areas and clearances; (b) proper ventilation and/or air conditioning; (c) adequate transportation facilities, such as elevators, stairways, and cranes or materials handling equipment; (d) adequate lighting; and (e) aesthetics.

The structural framework design is the selection of arrangement and sizes of structural elements so that service loads may be safely carried, and displacements are within acceptable limits.

The iterative design procedure may be outlined as follows:

1. Planning. Establishment of the functions for which the structure must serve. Set criteria against which to measure the resulting design for being an optimum.

2. Preliminary structural configuration. Arrangement of the elements to serve the functions in step 1.

3. Establishment of the loads to be carried.

4. Preliminary member selection. Based on the decisions of steps 1, 2, and 3 selection of the member sizes to satisfy an objective criterion, such as least weight or cost.

5. Analysis. Structural analysis involving modeling the loads and the structural framework to obtain internal forces and any desired deflections.

6. Evaluation. Are all strength and serviceability requirements satisfied and is the result optimum? Compare the result with predetermined criteria.

7. Redesign. Repetition of any part of the sequence 1 through 6 found necessary or desirable as a result of evaluation. Steps 1 through 6 represent an iterative process. Usually in this text only steps 3 through 6 will be subject to this iteration since the structural configuration and external loading will be prescribed.

8. Final decision. The determination of whether or not an optimum design has been achieved.

3.3 Historical Background of Steel Structures

Metal as a structural material began with cast iron, used on a 100ft (about. 30m) arch span which was built in England in 1777—1779. A number of cast-iron bridges were built during the period 1780—1820, mostly arch-shaped with main girders consisting of individual cast-iron pieces forming bars or trusses. Cast iron was also used for chain links on suspension bridges until about 1840.

Wrought iron began replacing cast iron soon after 1840, the earliest important example being the Brittania Bridge over Menai Straits in Wales, which was built in 1846—1850. This was a tubular girder bridge having spans 230-460-460-230 ft (about 70-140-140-70m), which was made from wrought-iron plates and angles.

The process of rolling various shapes was developing as cast iron and wrought iron received wider usage. Bars were rolled on an industrial scale beginning about 1780. The rolling of rails began about 1820 and was extended to I-shapes by the 1870s.

The development of the Bessemer process (1855), the introduction of a basic liner in the Bessemer converter (1870), and the open-hearth furnace brought widespread use of iron ore products in building materials. Since 1890, steel has replaced wrought iron as the principal metallic building material. Currently (1989), steels having yield stresses varying from 24000 to 100000 pounds per square inch, psi (165 to 690 megapascals, MPa), and available for structural uses.

3.4 Loads

The accurate determination of the loads to which a structure or structural element will be subjected is not always predictable.[2] Even if the loads are well known at one location in a structure, the distribution of load from element to element throughout the structure usually requires assumptions and approximations. Some of the most common kinds of loads are discussed in the following sections.

3.4.1 Dead Load

Dead load is a fixed-position gravity service load, so called because it acts continuously toward the earth when the structure is in service.[3] The weight of the structure is considered dead load, as well as attachments to the structure such as pipes, electrical conduit, air-conditioning and heating ducts, lighting fixtures, floor covering, roof covering, and suspended ceilings; that is, all items that remain throughout the life of the structure.

Dead loads are usually known accurately but not until the design has been completed. Under steps 3 through 6 of the design procedure, the weight of the structure or structural element must be estimated, preliminary section selected, weight recomputed, and member selection revised if necessary. The dead load of attachments is usually known with reasonable accuracy prior to the design.

3.4.2 Live Load

Gravity loads acting when the structure is in service, but varying in magnitude and location, are termed live loads.[4] Examples of live loads are human occupants, furniture, movable equipment, vehicles, and stored goods. Some live loads may be practically permanent, others may be highly transient. Because of the unknown nature of the magnitude, location, and density of live load items, realistic magnitudes and the positions of such loads are very difficult to determine.

Because of the public concern for adequate safety, live loads to be taken as service loads in design are usually prescribed by state and local building codes. These loads are generally empirical and conservative, based on experience and accepted practice rather than accurately computed values. Wherever local codes do not apply, or do not exist, the provisions from one of several regional and national building codes may be used. One such widely recognized code is the *American national Standard Minimum Design Loads for Buildings and Other Structures* ANSI A58. 1 of the American National Standards Institute (ANSI), from which some typical live loads are presented. The code will henceforth be referred to as the ANSI Standard. This standard is updated from time to time, most recently in 1982.

Live load when applied to a structure should be positioned to give the maximum effect, including partial loading, alternate span loading, or full span loading as may be necessary.[5] The simplified assumption of full uniform loading everywhere should be used only when it agrees with reality or is an appropriate approximation. The probability of having the prescribed loading applied uniformly over an entire floor, or over all floors of a building simultaneously, is almost nonexistent. Most codes recognize this by allowing for some percentage reduction from full loading. For instance, for live loads of 100 psf or more ANSI Standard allows members having an influence area of 400 ft^2 or more to be designed for a reduced live load according to Eq. 3-1, as follows:

$$L = L_0 \left(0.25 + \frac{15}{\sqrt{A_I}}\right) \tag{3-1}$$

where L—reduced live load per sq ft of area supported by the member;

L_0—unreduced live load per sq ft of area supported by the member (from Table 3-1);

A_I—influence area (ft^2)。

Table 3-1 Typical minimum uniformly distributed live loads

Occupancy or use	Live load	
	psf	Pa[①]
1.Hotel guest rooms	40	1900
School classrooms		
Private apartments		
Hospital private rooms		
2.Offices	50	2400
3.Assembly halls, fixed seat	60	2900
Library reading rooms		
4.Corridors, above first floor in schools.libraries and hospitals	80	3800
5.Assembly areas: theater lobbies	100	4800
Dining rooms and restaurants		
Office building lobbies		
Main floor retails stores		
Assembly hall movable seats		
6.Wholesale stores, all floors	125	6000
Manufacturing light		
Storage warehouses, light		
7.Armories and drill halls	150	7200

(Continued)

Occupancy or use	Live load	
	psf	Pa[①]
Stage floors		
Library stack rooms		
8. Manufacturing, heavy	250	12000
Sidewalks and driveways subject to trucking		
Storage warehouse, heavy		

① SI values are approximate conversions. 1psf ($1b/ft^2$) = 47.9Pa.

The influence area A_I is four times the tributary area (the area which distributes load to member being considered) for a column, two times the tributary area for a beam, and is equal to the panel area for a two-way slab. The reduced live load L shall not be less than 50% of the live load L_0 for members supporting one floor, nor less than 40% of the live load L_0 otherwise.

The live load reduction referred to above is not permitted in areas to be occupied as places of public assembly and for one-way slabs, when the live load L is 100 psf or less. Reductions are permitted for occupancies where L_0 is greater than 100 psf and for garages and roofs only under special circumstances (ANSI-4.7.2).

3.4.3 Highway Live Loads

Highway vehicle loading in the United States has been standardized by the American Association of State Highway and Transportation Officials (AASHTO) into standard truck loads and lane loads that approximate a series of trucks. There are two systems, designated H and HS, that are identified by the number of axles per truck. The H system has two axles, whereas the HS system has three axles per truck. There are several classes of loading; however, the usual ones are known as H20 and HS20, shown in Fig. 3-1.

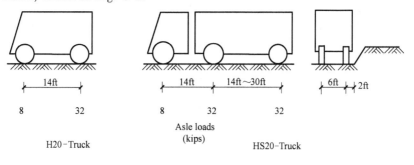

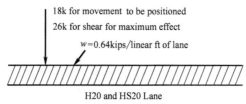

Fig. 3-1 AASHTO highway H20 and HS20 loading.

Note: 1kip = 4.45kN.

In designing a given bridge, either one truck loading is applied to the entire structure, or the lane loading is applied. When the lane loading is used, the uniform portion is distributed over as much of the span or spans as will cause the maximum effect. In addition, the one concentrated load (for maximum negative moment on continuous spans, a second concentrated load is also used) is positioned for the most severe loading effect. The load distribution across the width of a bridge to its various supporting members is taken in accordance with semiempirical rules that depend on the type of bridge deck and supporting structure.

The single truck loading provides the effect of a heavy concentrated load and usually governs on relatively short spans. The uniform lane load is to simulate a line of traffic, and the added concentrated load is to account for the possibility of one extra heavy vehicle in the line of traffic. These loads have been used with no apparent difficulty since 1944, before which time a line of trucks was actually used for the loading. On the interstate system of highways, a military loading is also used that consists of two 24kip (about 107kN) axle loads spaced 4ft (about 1.2m) apart.

Railroad bridges are designed to carry a similar semiempirical loading known as the Cooper E72 train, consisting of a series of concentrated loads a fixed distance apart followed by uniform loading. This loading is prescribed by the American Railway Engineering Association (AREA).

3.4.4 Impact

The term impact as ordinarily used in structural design refers to the dynamic effect of a suddenly applied load. In the building of a structure, the materials are added slowly; people entering a building are also considered a gradual loading. Dead loads are static loads; i.e., they have no effect other than weight. Live loads may be either static or they may have a dynamic effect. Persons and furniture would be treated as static live load, but cranes and various types of machinery also have dynamic effects.

Consider the spring-mass system of Fig. 3-2a, where the spring may be thought of as analogous to an elastic beam. When load is gradually applied (i.e., static loading) the mass (weight) deflects an amount x and the load on the spring (beam) is equal to the weight W. In Fig. 3-2b the load is suddenly applied (dynamic loading), and the maximum deflection is $2x$; i.e., the maximum load on the spring (beam) is $2W$. In this case the mass vibrates in simple harmonic motion with its neutral position equal to its static deflected position. In real structures, the harmonic (vibratory) motion is damped out (reduced to zero) very rapidly. Once the motion has stopped, the force remaining in the spring is the weight W. To account for the increased force during the time the member is in motion, a load equal to twice the static load W should be used—add 100% of the static load to represent the dynamic effect. This is called a 100% impact factor.

Any live load that can have a dynamic effect should be increased by an impact factor. While a dynamic analysis of a structure could be made, such a procedure is unnecessary in ordinary design. Thus empirical formulas and impact factors are usually used. In cases where the dynamic effect is small (say where impact would be less than about 20%), it is ordinarily accounted for by using a conservative (higher) value for the specified live load. The dynamic effects of persons in buildings

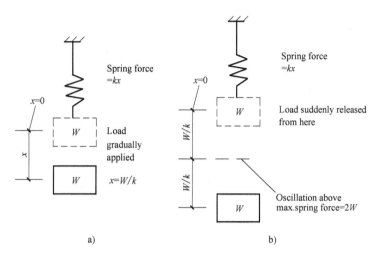

Fig. 3-2 Comparison of static and dynamic loading.
a) No vibration max. spring force = W b) Free vibration max. Spring force = $2W$

and of slow-moving vehicles in parking garages are examples where ordinary design live load is conservative, and usually no explicit impact factor is added.

For highway bridge design, however, impact is always to be considered. AASHTO prescribes empirically that the impact factor expressed as a portion of live load is

$$I = 50/(L+125) \leqslant 0.30 \qquad (3\text{-}2)$$

In Eq. 3-2, L (expressed in feet) is the length of the portion of the span that is loaded to give the maximum effect on the member. Since vehicles travel directly on the superstructure, all parts of it are subjected to vibration and must be designed to include impact. The substructure, including all portions not rigidly attached to the superstructure such as abutments, retaining walls, and piers, are assumed to have adequate damping or be sufficiently remote from the application point of the dynamic load so that impact might not be considered. Again, conservative static loads may account for the smaller dynamic effects.

In buildings, it is principally in the design of supports for cranes and heavy machinery that impact is explicitly considered. The American Institute of Steel Construction (AISC) Allowable Stress Design (ASD) and Load and Resistance Factor Design (LRFD) Specifications (ASD and LRFD-A4.2) state that if not otherwise specified, the impact percentage shall be:

For supports of elevators and elevator machinery, 100%.

For supports of light machinery, shaft or motor driven, not less than, 20%.

For supports of reciprocating machinery or power driven units, not less than, 50%.

For hangers supporting floors and balconies, 33%.

For cab-operated traveling crane support girders and their connections, 25%.

For pendant-operated traveling crane support girders and their connections, 10%.

In the design of crane runway beams and their connections (Fig. 3-3), the horizontal forces caused by moving crane trolleys must be considered. Both LFRD and ASD-A4.3 prescribe using a minimum of "20% of the sum of the lifted load and the crane trolley (but exclusive of other parts of the crane). The force shall be assumed to be applied at the top of the rails, acting in either direction normal to the runway rails, and shall be distributed with due regard for lateral stiffness of the structure supporting the rails."

In addition, due to acceleration and deceleration of the entire crane, a longitudinal tractive force is transmitted to the runway girder through friction of the end truck wheels with the crane rail.[6] LRFD and ASD-A 4. 3 require this force, unless otherwise specified, to be a minimum of 10% of the maximum wheel loads of the crane applied at the top of the rail.

The reader will find continued reference to the AISC Specifications (ASD and LRFD) which are contained, respectively in the *AISC ASD Manual* and *AISC LRFD Manual*. These two books may be purchased from AISC, 400 North Michigan Avenue, Chicago, IL 60611-4185.

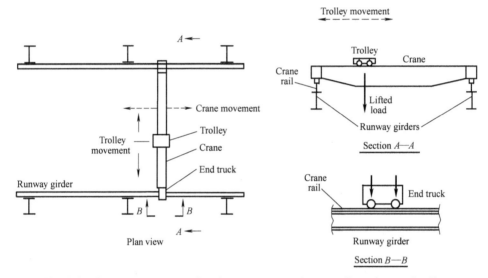

Fig. 3-3 Crane arrangement showing movements that contribute impact loading.

3.4.5 Snow Load

The live loading for which roofs are designed is either totally or primarily a snow load. Since snow has a variable specific gravity, even if one knows the depth of snow for which design is to be made, the load per unit area of roof is at best only a guess.

The best procedure for establishing snow load for design is to follow the ANSI Standard. This Code uses a map of the United States giving isolines of ground snow corresponding to a 50-year mean recurrence interval for use in designing most permanent structures. The ground snow is then multiplied by a coefficient that includes the effect of roof slope, wind exposure, nonuniform accumulation on pitched or curved roofs, multiple series roofs, and multilevel roofs and roof areas adjacent to projections on a roof level.

It is apparent that the steeper the roof the less snow can accumulate. Also, partial snow loading must be considered in addition to full loading, if it is believed such loading can occur and would cause maximum effects. Wind may also act on a structure that is carrying snow load. It is unlikely, however, that maximum snow and wind loads would act simultaneously.

In general, the basic snow load used in design varies from 30 to 40psf (1400 to 1900MPa) in the northern and eastern states to 20psf (960MPa) or less in the southern states. Flat roofs in normally warm climates should be designed for 20psf (960MPa) even when such accumulation of snow may seem doubtful. This loading may be thought of as due to people gathered on such a roof. Furthermore, though wind is frequently ignored as a vertical force on a roof, nevertheless it may cause such an effect. For these reasons, a 20 psf (960MPa) minimum loading, even though it may not always be snow, is reasonable. Local codes, actual weather conditions, ANSI, or the *Canadian Structural Design Manual*, should be used when designing for snow.

Other snow load information has been provided in the *Building Structural Design Handbook* and related studies.

3.4.6 Wind Load

All structures are subject to wind load, but it is usually only those more than three or four stories high, other than long bridges, for which special consideration of wind is required.[7]

On any typical building of rectangular plan and elevation, wind exerts pressure on the windward side and suction on the leeward side, as well as either uplift or downward pressure on the roof. For most ordinary situations, vertical roof loading from wind is neglected on the assumption that snow loading will require a greater strength than wind loading. This assumption is not true for southern climates where the vertical loading due to wind must be included. Furthermore, the total lateral wind load, windward and leeward effect, is commonly assumed to be applied to the windward face of the building.

In accordance with Bernoulli's theorem for an ideal fluid striking an object, the increase in static pressure equals the decrease in dynamic pressure, or

$$q = \frac{1}{2}\rho V^2 \tag{3-3}$$

Where q is the dynamic pressure on the object, ρ is the mass density of air [specific weight $w = 0.07651$ pcf (1pcf = 16.02kg/m^3) at sea level and 15℃], and V is the wind velocity. In terms of velocity V in miles per hour, the dynamic pressure q (psf) would be

$$q = \frac{1}{2}\left(\frac{0.07651}{32.2}\right)\left(\frac{5280V^2}{3600}\right) = 0.0026V^2 \tag{3-4}$$

In design of usual types of buildings, the dynamic pressure q is commonly converted into equivalent static pressure p, which may be expressed

$$p = qC_e C_g C_p \tag{3-5}$$

Where C_e is an exposure factor that varies from 1.0 (for 0~40ft height) to 2.0 (for 740~1200ft height); C_g is a gust factor, such as 2.0 for structural members and 2.5 for small elements including cladding; and C_p is a shape factor for the building as a whole. Excellent details of application of wind loading to structures are available in the ANSI Standard and in the *National Building Code of*

Canada.

The commonly used wind pressure of 20 psf, as specified by many building codes, corresponds to a velocity of 88 miles per hour (mph) from Eq. 3-4. An exposure factor C_e of 1.0, a gust factor C_g of 2.0, and a shape factor C_p of 1.3 for an airtight building, along with a 20 psf equivalent static pressure p, will give from Eq. 3-5 a dynamic pressure q of 7.7 psf, which corresponds using Eq. 3-4, to a wind velocity of 55 mph. For all buildings having nonplanar surfaces, plane surfaces inclined to the wind direction, or surfaces having significant openings, special determination of the wind forces should be made using such sources as the ANSI Standard, or the *National Building Code of Canada*. For more extensive treatment of wind loads, the reader is referred to the Task Committee on Wind Forces.

3.4.7 Earthquake Load

An earthquake consists of horizontal and vertical ground motions, with the vertical motion usually having much the smaller magnitude.[8] Because the horizontal motion of the ground causes the most significant effect, it is that effect which is usually thought of as earthquake load. When the ground under an object (structure) having a certain mass suddenly moves, the inertia of the mass tends to resist the movement, as shown in Fig. 3-4. A shear force is developed between the ground and the mass. Most building codes having earthquake provisions require the designer to consider a lateral force CW that is usually empirically prescribed. The dynamics of earthquake action on structures is outside the scope of this text, and the reader is referred to Chopra, Clough and Penzien.

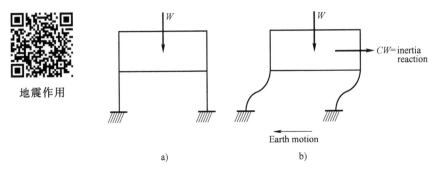

Fig. 3-4 Force developed by earthquake.

a) At test   b) Under horizontal motion from earthquake

In order to simplify the design process, most building codes contain an equivalent lateral force procedure for designing to resist earthquakes. One of the most widely used design recommendations is that of the Structural Engineers Association of California (SEAOC), the latest version of which is 1974. Since that time, the Applied Technology Council (ATC) prepared a set of design provisions. Some recent rules for the equivalent lateral force procedure are those given by the ANSI Standard. In ANSI the lateral seismic force V, expressed as follows, are assumed to act nonconcurrently in the direction of each of the main axes of the structure.

$$V = ZIKCSW \tag{3-6}$$

Where Z—seismic zone coefficient, varying from 1/8 for the zone of lowest seismicity, to 1 for the zone of highest seismicity;

I—occupancy importance factor, varying from 1.5 for buildings designated as "essential facilities", 1.25 for buildings where the primary occupancy is for assembly for greater than 300 persons, to 1.0 for usual buildings;

K—horizontal force factor, varying from 0.67 to 2.5, indicating capacity of the structure to absorb plastic deformation (low values indicate high ductility);

$$C = \frac{1}{15\sqrt{T}} \leq 0.12 \tag{3-7}$$

C—the seismic coefficient, equivalent to the maximum acceleration in term of acceleration due to gravity;

T—fundamental natural period, i.e., time for one cycle of vibration, of the building in the direction of motion;

S—soil profile coefficient, varying from 1.0 for rock to 1.5 for soft to medium-stiff clays and sands;

W—total dead load of the building, including interior partitions.

When the natural period T cannot be determined by a rational means from technical data, it may be obtained as follows for shear walls or exterior concrete frames utilizing deep beams or or wide piers, or both:

$$C = \frac{0.05 h_n}{\sqrt{D}} \tag{3-8}$$

where D is the dimension of the structure in the direction of the applied forces, in feet, and h_n is the height of the building.

Once the base shear V has been determined, the lateral force must be distributed over the height of the building.

More details of the ANSI Standard procedure are available in the *Building Structural Design Handbook*. Various building code formulas for earthquake-resistant design are compared by Chopra and Cruz. Many states have adopted the *Uniform Building Code* (UBC), the most recent version of which is 1985, which contains provisions for design to resist earthquakes generally based on the ANSI Standard.

3.5 Types of Structural Steel Members

The function of a structure is the principal factor determining the structural configuration. Using the structural configuration along with the design loads, individual components are selected to properly support and transmit loads throughout the structure. Steel members are selected from among the standard rolled shapes adopted by the American Institute of Steel Construction (AISC) (also given by American Society for Testing and Materials [ASTM] A6 Specification). Of course, welding permits combining plates and/or other rolled shapes to obtain any shape the designer may require.

Typical rolled shapes, the dimensions for which are found in the AISC Manual, are shown in Fig. 3-5. The most commonly used section is the wide-flange shape (Fig. 3-5a) which is formed by hot rolling in the steel mill. The wide-flange shape is designated by the nominal depth and the weight per foot, such as a W18×97 which is nominally 18 in. deep (actual depth = 18.59 in. according to AISC Manual) and weighs 97 pounds per foot (In SI units the W18×97 section could be designated W460×142, meaning nominally 460mm deep and having a mass of 142kg/m). Two sets of dimensions are found in the AISC Manual, one set stated in decimals for the designer to use in computations, and another set expressed in fractions (1/16 in. as the smallest increment) for the detailer to use on plans and shop drawings. Rolled W shapes are also designated by ANSI/ASTM A6 in accordance with web thickness as Groups I through V, with the thinnest web sections in Group 1.

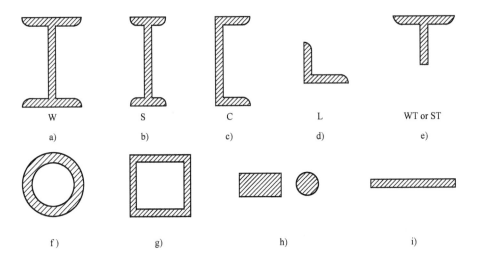

Fig. 3-5 Standard rolled shapes.
a) Wide-flange shape   b) American standard beam   c) American standard channel   d) Angle
e) Structural tee   f) Pipe section   g) Structural turbing   h) Bars   i) Plates

The American Standard beam (Fig. 3-5b), commonly called the I-beam, has relatively narrow and sloping flanges and a thick web compared to the wide-flange shape. Use of most I-beams has become relatively uncommon because of excessive material in the web and relative lack of lateral stiffness due to the narrow flanges.

The channel (Fig. 3-5c) and angle (Fig. 3-5d) are commonly used either alone or in combination with other sections. The channel is designated, for example, as C12×20.7, a nominal 12in. deep channel having a weight of 20.7 pounds per foot. Angles are designated by their leg length (long leg first) and thickness, such as, L6×4×3/8.

The structural tee (Fig. 3-5e) is made by cutting wide-flange or I-beams in half and is commonly used for chord members in trusses. The tee is designated, for example, as WT5×44, where the 5 is the nominal depth and 44 is the weight in pounds per foot; this tee being cut from a W10×88.

Pipe sections (Fig. 3-5f) are designated "standard" "extra strong" and "double-extra strong" in accordance with the thickness and are also nominally prescribed by diameter; thus 10in.-diam double-extra strong is an example of a particular pipe size.

Structural tubing (Fig. 3-5g) is used where pleasing architectural appearance is desired with exposed steel. Tubing is designated by outside dimensions and thickness, such as structural tubing, 8×6×1/4.

The sections shown in Fig. 3-5 are all hot-rolled; that is, they are formed from hot billet steel (blocks of steel) by passing through rolls numerous times to obtain the final shapes.

Many other shapes are cold-formed from plate material having a thickness not exceeding 1 in., as shown in Fig. 3-6.

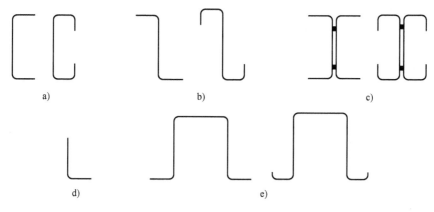

Fig. 3-6 Some cold-formed shapes.
a) Channels   b) Zees   c) I-shaped double channels   d) Angle   e) Hat sections

Regarding size and designation of cold-formed steel members, there are no truly standard shapes even though the properties of many common shapes are given in the *Cold-Formed Steel Design Manual*. Various manufacturers produce many proprietary shapes.

3.5.1 Tension Members

The tension member occurs commonly as a chord member in a truss, as diagonal bracing in many types of structures, as direct support for balconies, as cables in suspended roof systems, and as suspension bridge main cables and suspenders that support the roadway. Typical cross-sections of tension members are shown in Fig. 3-7.

3.5.2 Compression Members

Since compression member strength is a function of the cross-sectional shape (radius of gyration), the area is generally spread out as much as is practical. Chord members in trusses, and many interior columns in buildings are examples of members subject to axial compression. Even under the most ideal condition, pure axial compression is not attainable; so, design for "axial" loading assumes the effect of any small simultaneous bending may be neglected. Typical cross-sections of com-

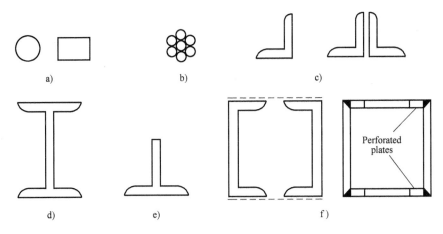

Fig. 3-7 Typical tension members.

a) Round and rectangular bars, including eye bars and upset bars   b) Cables composed of many small wires
c) Single and double angles   d) Rolled W-and S-sections   e) Structural tee   f) Built-up box sections

pression members are shown in Fig. 3-8.

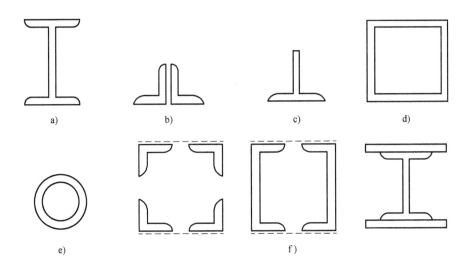

Fig. 3-8 Typical Compression members.

a) Rooled W-and S-shapes   b) Double angle   c) Structural tee
d) Structural tubing   e) Pipe section   f) Built-up sections

3.5.3 Beams

Beams are members subjected to transverse loading and are most efficient when their area is distributed so as to be located at the greatest practical distance from the neutral axis.[9] The most common beam sections are the wide-flange (W) and I-beams (S) (Fig. 3-9a), as well as smaller rolled I-shaped sections designated as "miscellaneous shapes" (M).

For deeper and thinner-webbed sections than can economically be rolled, welded I-shaped sections (Fig. 3-9b) are used, including stiffened plate girders.

For moderate spans carrying light loads, open-web "joists" are often used (Fig. 3-9c). These are parallel chord truss-type members used for the support of floors and roofs. The steel may be hot-rolled or cold-formed. Such joists are designated "K-Series" "LH-Series" and "DLH-Series". The K-Series is suitable for members having the direct support of floors and roof decks in buildings. The LH-Series and DLH-Series are known as Longspan and Deep Longspan, respectively. Longspan Steel Joists are shop-fabricated trusses used "for the direct support of floor or roof slabs or decks between walls, beams, and main structural members". Deep Longspan Joists are used " for the direct support of roof slabs or decks between wall, beams and main structural members". The design of the chords for K-Series trusses is based on a yield strength of 50ksi (about 345MPa), while the web sections may use either 36ksi (about 248MPa) or 50ksi (about 345MPa). For the LH-and DLH-Series the chord and web sections design must be based on a yield strength of at least 36 ksi (about 248MPa) but not greater than 50ksi (about 345MPa).

The K-Series joists have depths from 8 to 30 in. for clear spans to 60 ft. The Longspan joists (LH-Series) have depths from 18 to 48 in. for clear spans to 96 ft. The Deep Longspan joists (DLH-Series) have depths from 52 to 72 in. for clear spans to 144 ft.

All of these joists are designed according to Specifications adopted by the Steel Joist Institute (SJI), which generally are in agreement with the AISC Specifications for hot-rolled steels and AISI Specifications for cold-formed steels.

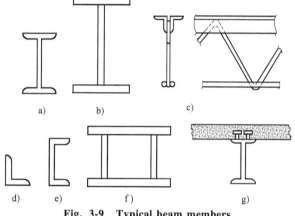

Fig. 3-9 Typical beam members.
a) Rooled W-and other I-shaped sections   b) Welded I-shape (plate girder)   c) Open web joists   d) Angle   e) Channel
f) Built-up members   g) Composite steel-concrete

For beams (known as lintels) carrying loads across window and door openings, angles (Fig. 3-9d) are frequently used; and for beams (known as girts) in wall panels, channels (Fig. 3-9e) are frequently used.

3.5.4 Bending and Axial Load

When simultaneous action of tension or compression along with bending occurs, a combined stress problem arises and the type of member used will be dependent on the type of stress that predominates.[10]

The aforementioned illustration of types of members to resist various kinds of stress is intended only to show common and representative types of members and not to be all inclusive.

New Words and Expressions

ventilation   *n.*   通风换气设备
cast iron   铸铁
wrought iron   熟铁，锻铁
Bessemer process   酸性转炉法
converter   *n.*   炼钢炉，吹风转炉
tractive   *adj.*   牵引的
isoline   *n.*   等值线，等位线
lee   *n.*   背风处，下风
windward   *n.*   迎风面
partition   *n.*   隔墙，分割
radius of gyration   回转半径
miscellaneous   *adj.*   混杂的，多方面的，有各种特点的
channel   *n.*   槽钢，槽，沟
lintel   *n.*   过梁

Notes

1. Structural design may be defined as a mixture of art and science, combining the experienced engineer's intuitive feeling for the behavior of a structure with a sound knowledge of the principles of statics, dynamics, mechanics of materials, and structural analysis, to produce a safe economical structure that will serve its intended purpose.

本句难点解析：句子中 combining... analysis 作为定语修饰 structural design，to produce a safe economical structure that will serve its intended purpose 则为状语从句，表示目的。

本句大意如下：结构设计可以被定义为一种艺术与科学的结合体，它将有经验的工程师对建筑性能的直觉，与静力、动力、材料力学以及结构分析的相关知识联合起来，产生了一个可以实现其既定目标的安全经济的结构。

2. The accurate determination of the loads to which a structure or structural element will be subjected is not always predictable.

本句难点解析：句子的主体是 The accurate determination of the loads is not always predictable。其中 to which a structure or structural element will be subjected 作为定语从句修饰 loads。

本句大意如下：结构或者结构单元的受力通常是不可精确预测的。

3. Dead load is a fixed-position gravity service load, so called because it acts continuously toward the earth when the structure is in service.

本句难点解析：句子中 so called 表示"之所以这么称呼"，起转折连接作用。

本句大意如下：恒荷载是一个固定位置的重力荷载，之所以这么称呼，是因为在结构使用阶段其作用方向一直指向地面。

Unit 3 Introduction to Steel Structures

4. Gravity loads acting when the structure is in service, but varying in magnitude and location, are termed live loads.

本句难点解析：句子的主体是 Gravity loads are termed live loads。其中 acting when the structure is in service, but varying in magnitude and location 作为定语修饰限定 gravity loads，表明其范围。

本句大意如下：在结构使用期间作用于其上，但大小以及位置变化的重力荷载，称为活荷载。

5. Live load when applied to a structure should be positioned to give the maximum effect, including partial loading, alternate span loading, or full span loading as may be necessary.

本句难点解析：句子的主体是 Live load should be positioned to give the maximum effect。

本句大意如下：当活荷载施加于结构时，应该将其放置在产生最大效应的位置上，包括部分加载、分跨加载或者全跨加载（若有需要）。

6. In addition, due to acceleration and deceleration of the entire crane, a longitudinal tractive force is transmitted to the runway girder through friction of the end truck wheels with the crane rail.

本句难点解析：其中 through 作为介词表明了牵引力的传导方式。

本句大意如下：此外，由于整个起重机的加速以及减速，纵向的牵引力将通过末端车轮与起重机轨道之间的摩擦力进行传递。

7. All structures are subject to wind load, but it is usually only those more than three or four stories high, other than long bridges, for which special consideration of wind is required.

本句难点解析：句子中 for which special consideration of wind is required. 作为定语从句修饰前句。

本句大意如下：所有的结构均承受风荷载，但除了长大桥以外，通常仅有高于三四层的建筑需要对风荷载进行专门考虑。

8. An earthquake consists of horizontal and vertical ground motions, with the vertical motion usually having much the smaller magnitude.

本句难点解析：句子中 with the vertical motion usually having much the smaller magnitude 为伴随状语。进一步解释说明竖向运动的情况。

本句大意如下：地震包含水平以及竖向运动，其中竖向运动幅度较小。

9. Beams are members subjected to transverse loading and are most efficient when their area is distributed so as to be located at the greatest practical distance from the neutral axis.

本句难点解析：句子中 subjected to transverse loading 作为定语修饰 members。so...axis 则是对 distributed 的进一步说明。

本句大意如下：梁是用于承受横向荷载的构件，且当它们面积分散，位于离轴心处实际距离最远的位置时最高效。

10. When simultaneous action of tension or compression along with bending occurs, a combined stress problem arises and the type of member used will be dependent on the type of stress that predominates.

本句难点解析：句子中 along with 表示同时发生，即拉、压、弯同时作用。

本句大意如下：当同时作用拉、压、弯时，将产生复合应力问题，此时构件类型的选取将取决于哪种应力起主要作用。

Exercises

Translate the following phrases into Chinese.

1. structural design
2. computations involving scientific principles
3. measurable criteria
4. adequate transportation facilities
5. preliminary structural configuration
6. yield stress
7. require assumptions and approximations
8. fixed-position gravity service load
9. ANSI Standard
10. maximum negative moment

Translate the following Sentences into Chinese.

1. If a specific objective criterion can be expressed mathematically, then optimization techniques may be employed to obtain a maximum or minimum for the objective function. Optimization procedures and techniques comprise an entire subject that is outside the scope of this text. The criterion of minimum weight is emphasized throughout, under the general assumption that minimum material represents minimum cost.

2. The structural framework design is the selection of arrangement and sizes of structural elements so that service loads may be safely carried, and displacements are within acceptable limits.

3. The process of rolling various shapes was developing as cast iron and wrought iron received wider usage.

4. Because of the public concern for adequate safety, live loads to be taken as service loads in design are usually prescribed by state and local building codes.

5. When the ground under an object (structure) having a certain mass suddenly moves, the inertia of the mass tends to resist the movement.

Unit 4

Seismic Design

▰ 4.1 Introduction

Earthquakes result from the sudden movement of tectonic plates in the earth's crust. The movement takes place at fault lines, and the energy released is transmitted through the earth in the form of waves that cause ground motion many miles from the epicenter.[1] Regions adjacent to active fault lines are the most prone to experience earthquakes. The values, expressed as a percent of gravity, represent the expected peak acceleration of a single-degree-of-freedom system with a 0.2 sec period and 5 percent of critical damping.[2] Known as the 0.2 sec spectral response acceleration S_s (subscript s for short period), it is used, along with the 1.0 sec spectral response acceleration S_1 (mapped in a similar manner), to establish the loading criteria for seismic design. Accelerations S_s and S_1 are based on historical records and local geology. For most of the country, they represent earthquake ground motion with a "likelihood of exceedance of 2 percent in 50 years", a value that is equivalent to a return period of about 2500 years.

As experienced by structures, earthquakes consist of random horizontal and vertical movements of the earth's surface. As the ground moves, inertia tends to keep structures in place (Fig. 4-1), resulting in the imposition of displacements and forces that can have catastrophic results.[3] The purpose of seismic design is to proportion structures so that they can withstand the displacements and the forces induced by the ground motion.

Historically in North America, seismic design has emphasized the effects of horizontal ground motion because the horizontal components of an earthquake usually exceed the vertical components and because structures are usually much stiffer and stronger in response to vertical loads than they are in response to horizontal loads. Experience has shown that the horizontal components are the most destructive. For structural design, the intensity of an earthquake is usually described in terms of the peak ground acceleration as a fraction of the acceleration of gravity, i.e., $0.1g$, $0.2g$, or $0.3g$. Although peak acceleration is an important design parameter, the frequency characteristics and duration of an earthquake are also important; the closer the frequency of the earthquake motion is to the natural frequency of a structure and the longer the duration of the earthquake, the greater the potential for damage.

Based on elastic behavior, structures subjected to a major earthquake would be required to undergo large displacements. However, North American practice requires that structures be designed for only a fraction of the forces associated with those displacements. The relatively low design forces are justified by the observations that buildings designed for low forces have behaved satisfactorily and those structures dissipate significant energy as the materials yield

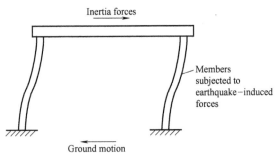

Fig. 4-1 Structure subjected to ground motion.

and behave inelastically. This nonlinear behavior, however, usually translates into increased displacements, which may require significant ductility and result in major nonstructural damage. Displacements may also be of such a magnitude that the strength of the structure is affected by stability considerations.

Designers of structures that may be subjected to earthquakes, therefore, are faced with a choice: (1) providing adequate stiffness and strength to limit the response of structures to the elastic range or (2) providing lower-strength structures, with presumably lower initial costs, that have the ability to withstand large inelastic deformations while maintaining their load-carrying capability.

4.2 Structural Response

The safety of a structure subjected to seismic loading rests on the designer's understanding of the response of the structure to ground motion. For many years, the goal of earthquake design in North America has been to construct buildings that will withstand moderate earthquakes without damage and severe earthquakes without collapse. Building codes have undergone regular modification as major earthquakes have exposed weaknesses in existing design criteria.

Design for earthquakes differs from design for gravity and wind loads in the relatively greater sensitivity of earthquake-induced forces to the geometry of the structure. Without careful design, forces and displacements can be concentrated in portions of a structure that are not capable of providing adequate strength or ductility. Steps to strengthen a member for one type of loading may actually increase the forces in the member and change the mode of failure from ductile to brittle.

结构响应

4.2.1 Structural Considerations

The closer the frequency of the ground motion is to one of the natural frequencies of a structure, the greater the likelihood of the structure experiencing resonance, resulting in an increase in both displacement and damage.[4] Therefore, earthquake response depends strongly on the geometric properties of a structure, especially height. Tall buildings respond more strongly to long-period (low-frequency) ground motion, while short buildings respond more strongly to short-period (high-frequency) ground motion. Fig. 4-2 shows the shapes for the principal modes of vibration of a three-story frame structure. The relative contribution of each mode to the lateral displacement of the structure depends on the frequency characteristics of the ground motion. The first mode (Fig. 4-2a) usually provides the greatest contribution to lateral displacement.

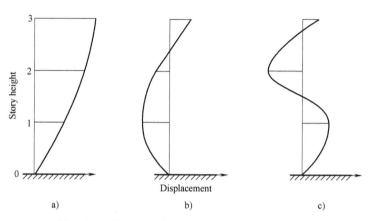

Fig. 4-2　Modal shapes for a three-story building.

a) First mode　b) Second mode　c) Third mode.

The taller a structure, the more susceptible it is to the effects of higher modes of vibration, which are generally additive to the effects of the lower modes and tend to have the greatest influence on the upper stories.[5] Under any circumstances, the longer the duration of an earthquake, the greater the potential for damage.

The configuration of a structure also has a major effect on its response to an earthquake. Structures with a discontinuity in stiffness or geometry can be subjected to undesirably high displacements or forces. For example, the discontinuance of shear walls, infill walls, or even cladding at a particular story level, such as shown in Fig. 4-3, will have the result of concentrating the displacement in the open, or "soft", story. The high displacement will, in turn, require a large amount of ductility if the structure is not to fail. Such a design is not recommended, and the Fig. 4-4 illustrates structures with vertical geometric and plan irregularities, which result in torsion induced by ground motion.

Within a structure, stiffer members tend to pick up a greater portion of the load. When a frame is combined with a shear wall, this can have the positive effect of reducing the displacements of the structure and decreasing both structural and non-structural damage. However, when the effects of higher stiffness members, such as masonry infill walls, are not considered in the design, unexpected and often undesirable results can occur.

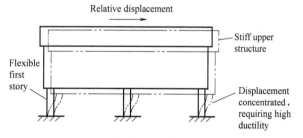

Fig. 4-3　Soft first story supporting a stiff upper structure.

Finally, any discussion of structural considerations would be incomplete without emphasizing the need to provide adequate separation between structures. Lateral displacements can result in structures coming in contact during an earthquake, resulting in major damage due to hammering.

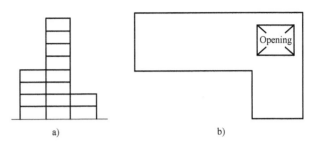

Fig. 4-4 Structures with vertical geometric and plan irregularities.
a) Vertical geometric b) Plan irregularities.

4.2.2 Member Considerations

Members designed for seismic loading must perform in a ductile fashion and dissipate energy in a manner that does not compromise the strength of the structure.[6] Both the overall design and the structural details must be considered to meet this goal.

The principal method of ensuring ductility in members subject to shear and bending is to provide confinement for the concrete. This is accomplished through the use of closed hoops or spiral reinforcement, which enclose the core of beams and columns. When confinement is provided, beams and columns can undergo nonlinear cyclic bending while maintaining their flexural strength and without deteriorating due to diagonal tension cracking. The formation of ductile hinges allows reinforced concrete frames to dissipate energy.

Successful seismic design of frames requires that the structures be proportioned so that hinges occur at locations that least compromise strength.[7] For a frame undergoing lateral displacement, such as shown in Fig. 4-5a, the flexural capacity of the members at a joint (Fig. 4-5b) should be such that the columns are stronger than the beams. In this way, hinges will form in the beams rather than the columns, minimizing the portion of the structure affected by nonlinear behavior and maintaining the overall vertical load capacity. For these reasons, the "weak beam-strong column" approach is used to design reinforced concrete frames subject to seismic loading.

When hinges form in a beam, or in extreme cases within a column, the moments at the end of the member, which are governed by flexural strength, determine the shear that must be carried, as illustrated in Fig. 4-5c. The shear V corresponding to a flexural failure at both ends of a beam or column is

$$V = \frac{M^+ + M^-}{l_n} \tag{4-1}$$

where M^+ and M^-—flexural capacities at the ends of the member;

l_n—clear span between supports.

The member must be checked for adequacy under the shear V in addition to shear resulting from dead and live gravity loads. Transverse reinforcement is added, as required. For members with inadequate shear capacity, the response will be dominated by the formation of diagonal cracks, rather than ductile hinges, resulting in a substantial reduction in the energy dissipation capacity of the member.[8]

抗震构件考虑

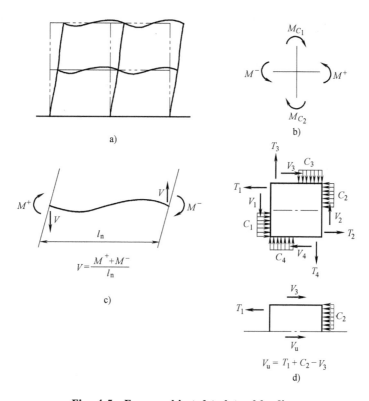

Fig. 4-5 Frame subjected to lateral loading.

a) Deflected shape b) Moments acting on beam-column joint c) Deflected shape and forces acting on a beam d) Forces acting on faces of a joint due to lateral load

If short members are used in a frame, the members may be unintentionally strong in flexure compared to their shear capacity. An example would be columns in a structure with deep spandrel beams or with "nonstructural" walls with openings that expose a portion of the columns to the full lateral load. As a result, the exposed region, called a captive column, responds by undergoing a shear failure, as shown in Fig. 4-6.

The lateral displacement of a frame places beam-column joints under high shear stresses because of the change from positive to negative bending in the flexural members from one side of the joint to the other, as shown in Fig. 4-5d. The joint must be able to withstand the high shear stresses and allow for a change in bar stress

Fig. 4-6 Shear failure in a captive column without adequate transverse reinforcement.

from tension to compression between the faces of the joint. Such a transfer of shear and bond is often made difficult by congestion of reinforcement through the joint. Thus, designers must ensure that joints not only have adequate strength but are also constructable. Two-way systems without beams are especially vulnerable because of low ductility at the slab-column intersection.

4.3 Seismic Loading Criteria

In the United States, the design criteria for earthquake loading are based on design procedures developed by the Building Seismic Safety Council and incorporated in *Minimum Design Loads for Buildings and Other Structures* (ASCE/SEI 7). The values of the spectral response accelerations S_S and $S1$, are obtained from detailed maps produced by the United States Geological Survey and included in ASCE/SEI 7. The values of S_S and $S1$ are used to determine the spectral response accelerations S_{Ds} and S_{D1} that are used in design.

$$S_{Ds} = \frac{2}{3} F_a S_s \tag{4-2}$$

$$S_{D1} = \frac{2}{3} F_a S_1 \tag{4-3}$$

where F_a and F_v are site coefficients that range from 0.8 to 0.25 and from 0.8 to 0.35, respectively, as a function of the geotechnical properties of the building site and the values of S_s and S_1, respectively. Higher values of F_a and F_v are possible for some sites. The coefficients F_a and F_v increase in magnitude as site conditions change from hard rock to thick, soft clays and (for softer foundations) as the values of S_s and S_1 decrease.

Both S_{Ds} and S_{D1} are used to construct the design response spectrum shown in Fig. 4-7, which relates the spectral response acceleration S_a, used to calculate the earthquake force, to the fundamental period of the structure T. In the spectrum, $T_0 = 0.2 S_{D1}/S_{Ds}$, $T_s = S_{D1}/S_{Ds}$, and T_L is the site-specific long-period transition period.

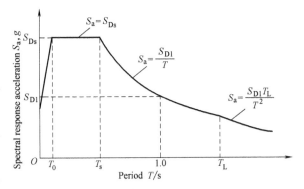

Fig. 4-7 Design response spectrum.

Structures are assigned to one of six *Seismic Design Categories* (SDCs) A through F as a function of (1) structure occupancy and use and (2) the values of S_{Ds} and S_{D1}. Requirements for seismic design and detailing are minimal for SDCs A and B but become progressively more rigorous for SDCs C through F.

Earthquake loading is included in two combinations of factored load.

$$U = 1.2D + 1.0E + 1.0L + 0.2S \tag{4-4}$$

$$U = 0.9D + 1.0E + 1.6H \tag{4-5}$$

where D—dead load;

E—earthquake load;
H—weight or pressure from soil;
L—live load;
S—snow load.

For SDC A, the earthquake load E is a horizontal load equal to 1 percent of the dead load D assigned to each floor. For SDCs B through F, the values of the earthquake load E used in Eq. (4-4) and Eq. (4-5) are, respectively,

$$E = \rho Q_E + 0.2 S_{Ds} D \tag{4-6a}$$
$$E = \rho Q_E - 0.2 S_{Ds} D \tag{4-6b}$$

where Q_E—effect of horizontal seismic forces;
ρ—reliability factor.

The factor ρ is taken as 1.0 for structures assigned to SDCs B and C and as 1.3 for structures assigned to SDCs D though F, in which case ρ may be taken as 1.0.

Combining Eq. (4-4) with Eq. (4-6a) and Eq. (4-5) with Eq. (4-6b) gives

$$U = (1.2 + 0.2 S_{Ds}) D + \rho Q_E + 1.0 L + 0.2 S \tag{4-7}$$
$$U = (0.9 - 0.2 S_{Ds}) D + \rho Q_E + 1.6 H \tag{4-8}$$

Eq. (4-4) and Eq. (4-6a) are used when dead load adds to the effects of horizontal ground motion, while Eq. (4-5) and Eq. (4-6b) are used when dead load counteracts the effects of horizontal ground motion. Thus, the total load factor for dead load is greater than 1.2 in Eq. (4-7) and less than 0.9 in Eq. (4-8).

ASCE/SEI 7 specifies six procedures (if SDC A is included) for determining the horizontal earthquake load Q_E. These procedures include three progressively more detailed methods that represent earthquake loading through the use of equivalent static lateral loads, *modal response spectrum analysis*, *linear time-history analysis*, and *nonlinear time-history analysis*. The method selected depends on the seismic design category. All but the most basic reinforced concrete structures in Seismic Design Categories B through F must be designed using equivalent lateral force analysis (the most detailed of the three equivalent static lateral load procedures), modal response analysis, or time-history analysis. These procedures are discussed next.

4.3.1 Equivalent Lateral Force Procedure

According to ASCE/SEI 7, equivalent lateral force analysis may be applied to all structures with S_{Ds} less than $0.33g$ and S_{D1} less than $0.133g$, as well as structures subjected to much higher design spectral response accelerations, if the structures meet certain requirements. More sophisticated dynamic analysis procedures must be used otherwise.

The equivalent lateral force procedure provides for the calculation of the total lateral force, defined as the design base shear V, which is then distributed over the height of the building. The design base shear V is calculated for a given direction of loading according to the equation

$$V = C_s W \tag{4-9}$$

where W is the total dead load plus applicable portions of other loads and

$$C_s = \frac{S_{Ds}}{R/I} \tag{4-10}$$

which need not be greater than

$$C_s = \frac{S_{D1}}{T(R/I)} \quad \text{for } T \leqslant T_L \tag{4-11}$$

or

$$C_s = \frac{S_{D1} T_L}{T^2 (R/I)} \quad \text{for } T \leqslant T_L \tag{4-12}$$

but may not be less than

$$C_s = 0.44 I S_{Ds} \geqslant 0.01 \tag{4-13}$$

or where $S_1 \geqslant 0.6g$,

$$C_s = \frac{0.5 S_1}{R/I} \tag{4-14}$$

where R—response modification factor (depends on structural system), Values of R for most reinforced concrete structures range from 4 to 8, based on ability of structural system to sustain earthquake loading and to dissipate energy;

I—occupancy important factor, $I = 1.0$, 1.25, or 1.5, depending upon the occupancy and use of structure;

T—fundamental period of structure.

According to ASCE/SEI 7, the period T can be calculated based on an analysis that accounts for the structural properties and deformational characteristics of the elements within the structure. Approximate methods may also be used in which the fundamental period of the structure may be calculated as

$$T = C_t h_n^x \tag{4-15}$$

where h_n—height above the base to the highest level of structure (ft);

C_t—0.016 for reinforced concrete moment-resisting frames in which frames resist 100 percent of required seismic force and are not enclosed or adjoined by more rigid components that will prevent frame from deflecting when subjected to seismic forces, and 0.020 for all other reinforced concrete buildings;

x—0.90 for $C_t = 0.016$ and 0.75 for $C_t = 0.020$.

Alternately, for structures not exceeding 12 stories in height, in which the lateral-force-resisting system consists of a moment-resisting frame and the story height is at least 10 ft,

$$T = 0.1 N \tag{4-16}$$

where N—number of stories.

For shear wall structures, ASCE/SEI 7 permits T to be approximated as

$$T = \frac{0.0019}{\sqrt{C_w}} h_n \tag{4-17}$$

where

$$C_w = \frac{100}{A_B} \sum_{i=1}^{n} \left(\frac{h_n}{h_i}\right)^2 \frac{A_i}{1 + 0.83(h_i/D_i)^2} \qquad (4\text{-}18)$$

where A_B—base area of structure (ft^2);

A_i—area of shear wall (ft^2);

D_i—length of shear wall i (ft);

n—number of shear walls in building that are effective in resisting lateral forces in direction under consideration.

The total base shear V is distributed over the height of the structure in accordance with Eq. (4-19).

$$F_x = \frac{w_x h_x^k}{\sum_{i=1}^{n} w_i h_i^k} \qquad (4\text{-}19)$$

where F_x—lateral seismic force induced at level x;

w_x, w_i—portion of W at level x and level i, respectively;

h_x, h_i—height to level x and level i, respectively;

k—exponent related to structural period, $k = 1$ for $T \leqslant 0.5$ sec and $k = 2$ for $T \geqslant 2.5$ sec, for $0.5\text{sec} < T < 2.5\text{sec}$, k is determined by linear interpolation or set to a value of 2.

The design shear at any story V_x equals the sum of the forces F_x at and above that story. For a 10-story building with a uniform mass distribution over the height and $T = 1.0$ sec, the lateral forces and story shears are distributed as shown in Fig. 4-8.

At each level, V_x is distributed in proportion to the stiffness of the elements in the vertical lateral-force-resisting system. To account for unintentional building irregularities that may cause a horizontal torsional moment, a minimum 5 percent eccentricity must be applied if the vertical lateral-force-resisting systems are connected by a floor system that is rigid in its own plane.

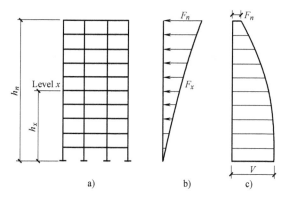

Fig. 4-8 Forces based on ASCE/SEI 7 equivalent lateral force procedure.

a) Structure b) Distribution of lateral forces over height c) Story shears

In addition to the criteria just described, ASCE/SEI 7 includes criteria to account for overturning effects and provides limits on story drift. $P\text{-}\Delta$ effects must be considered, and the effects of upward loads must be accounted for in the design of horizontal cantilever components and prestressed members.

4.3.2 Dynamic Lateral Force Procedures

ASCE/SEI 7 includes dynamic lateral force procedures that involve the use of (1) response spectra, which provide the earthquake-induced forces as a function of the natural periods of the structure, or (2) time-history analyses of the structural response based on a series of ground motion acceleration histories that are representative of ground motion expected at the site. Both procedures require the development of a mathematical model of the structure to represent the spatial distribution of mass and stiffness. Response spectra, are used to calculate peak forces for " a sufficient number of nodes to obtain the combined modal mass participation of at least 90 percent of the actual mass in each of two orthogonal directions". Since these forces do not always act in the same direction, as shown in Fig. 4-2, the peak forces are averaged statistically, in most cases using the square root of the sum of the squares to obtain equivalent static lateral forces for use in design. In cases where the periods in the translational and torsional modes are closely spaced and result in significant cross correlation of the modes, the complete quadratic combination method is used.9 When time-history analyses, which may include a linear or nonlinear representation of the structure, are used, design forces are obtained directly from the analyses.10 Both modal response spectrum and time-history procedures provide more realistic representations of the seismically induced forces in a structure than do equivalent lateral force analyses.

New Words and Expressions

    tectonic plates 构造板块，地壳板块
    fault lines 裂纹线，断层线
    critical damping 临界阻尼
    spectral response 谱响应
    subscript *n.* 下标，脚注
    dissipate *v.* 消散
    ductile to brittle 韧性到脆性
    discontinuity *n.* 不连续
    infill wall 填充墙
    incorporated in 纳入，包括
    cantilever *n.* 悬臂
    response spectra 反应谱
    orthogonal *adj.* 正交的

Notes

1. The movement takes place at fault lines, and the energy released is transmitted through the earth in the form of waves that cause ground motion many miles from the epicenter.

本句难点解析：句子中 through 作为介词表明能量传播途径，in the form of 表明传播方式，that cause ground motion many miles from the epicenter 作为定语从句修饰 waves。

本句大意如下：运动发生在断层线上，能量释放后以波的形式在地层传播，引起了离震源数英里之外的地方产生地面运动。

2. The values, expressed as a percent of gravity, represent the expected peak acceleration of a single-degree-of-freedom system with a 0.2sec period and 5 percent of critical damping.

本句难点解析：句子主体是 The values represent the expected peak acceleration。其中 with a 0.2sec period and 5 percent of critical damping 作为定语修饰 system，expressed as a percent of gravity 作为定语修饰 values。

本句大意如下：以重力百分数表示的值代表了在 0.2s 自振周期，临界阻尼为 0.05 的单自由度体系下的加速度峰值。

3. As the ground moves, inertia tends to keep structures in place (Fig. 4-1), resulting in the imposition of displacements and forces that can have catastrophic results.

本句难点解析：句子中 resulting in the imposition of displacements and forces that can have catastrophic results 作为定语从句表示结果。

本句大意如下：随着地面的运动，惯性将会使结构倾向于原地不动，从而在结构上施加了位移及力，这将导致严重的破坏后果。

4. The closer the frequency of the ground motion is to one of the natural frequencies of a structure, the greater the likelihood of the structure experiencing resonance, resulting in an increase in both displacement and damage.

本句难点解析：句子中 the closer...the greater 表示越接近……越大……。resulting in an increase in both displacement and damage 作为定语从句表示结果。

本句大意如下：地面运动的频率越接近结构自振频率，结构发生自振的可能性便越大，这将导致结构位移以及破坏情况的增加。

5. The taller a structure, the more susceptible it is to the effects of higher modes of vibration, which are generally additive to the effects of the lower modes and tend to have the greatest influence on the upper stories.

本句难点解析：句子中 the taller...the more 表示越高……越……。which 引导定语从句修饰 effects。

本句大意如下：结构越高，就越容易受高阶振型的影响，这种影响一般作为低阶模态作用的一个附加部分，且对上部楼层有较大的影响。

6. Members designed for seismic loading must perform in a ductile fashion and dissipate energy in a manner that does not compromise the strength of the structure.

本句难点解析：句子中 in a manner 表示以一种……的方式，并通过定语从句对 manner 进行说明。

本句大意如下：按地震荷载设计的构件必须以延性的方式工作，并以不会折减结构强度的方式进行耗能。

7. Successful seismic design of frames requires that the structures be proportioned so that hinges occur at locations that least compromise strength.

本句大意如下：成功的框架抗震设计要求结构比例合适，这样塑性铰才会出现在合适的位置，从而使得强度折减量最小。

8. For members with inadequate shear capacity, the response will be dominated by the formation of diagonal cracks, rather than ductile hinges, resulting in a substantial reduction in the energy dissipation capacity of the member.

本句难点解析：句子中 resulting in a substantial reduction in the energy dissipation capacity of the member 作为定语从句，表示目的。

本句大意如下：对于抗剪强度不足的构件，在地震作用下其主要先产生斜裂缝，而不是塑性铰，这将导致构件耗能能力的直接下降。

9. In cases where the periods in the translational and torsional modes are closely spaced and result in significant cross correlation of the modes, the complete quadratic combination method is used.

本句难点解析：句子中 In cases where 引导状语从句，其中 spaced 一词表示两者十分接近，即平移与扭转模态。

本句大意如下：在平移和扭转的模态周期比较接近，从而导致模态发生明显的交叉相关的情况下，可以采用完全二次型组合法。

10. When time-history analyses, which may include a linear or nonlinear representation of the structure, are used, design forces are obtained directly from the analyses.

本句难点解析：句子中 When 引导状语从句，从句主体是 When time-history analyses are used。which 引导定语从句修饰 time-history analyses。

本句大意如下：当使用时程分析，即包括结构的线性及非线性特征时，设计力可直接从分析中获得。

Translate the following phrases into Chinese.

1. earth's crust
2. a single-degree-of-freedom system
3. keep structures in place
4. mode of Vibration
5. lateral displacement
6. equivalent lateral force
7. fundamental period of the structure
8. non-structural damage
9. distribution of mass and stiffness
10. time-history procedure

Translate the following Sentences into Chinese.

1. Known as the 0.2 sec spectral response acceleration S_s (subscript s for short period), it is used, along withe the 1.0 sec spectral response acceleration S_1 (mapped in a similar manner), to establish the loading criteria for seismic design.

2. Although peak acceleration is an important design parameter, the frequency characteristics and duration of an earthquake are also important.

3. The configuration of a structure also has a major effect on its response to an earthquake.

4. According to ASCE/SEI 7, equivalent lateral force analysis may be applied to all structures with S_{Ds} less than $0.33g$ and S_{D1} less than $0.133g$, as well as structures subjected to much higher design spectral response accelerations, if the structures meet certain requirements.

5. The equivalent lateral force procedure provides for the calculation of the total lateral force, defined as the design base shear V, which is then distributed over the height of the building.

Unit 5

Composite Construction

5.1 Overview

In 1988, Bank of China Tower in Hong Kong (Fig. 5-1) was completed. This magnificent building was the tallest building in Asia at that time. The building was constructed using composite steel-concrete structures, which is an important milestone since the development of the composite construction.

Composite construction is a method of construction that is used in a variety of engineering and building applications. Employing dissimilar components, such as concrete and steel or fiberglass and foam, for a single use or structure is called composite construction. The goal is unifying of the individual component properties to create a composite material that possesses the desired properties of all component pieces.

In structural engineering, composite construction exists when two different materials are bound together so strongly that they act together as a single unit from a structural point of view. When this occurs, it is called composite action. One common example involves steel beams supporting concrete floor slabs. If the beam is not connected firmly to the slab, then the slab transfers all of its weight to the beam and the slab contributes nothing to the load carrying capability of the beam.[1] However, if the slab is connected positively to the beam with studs, then a portion of the slab can be assumed to act compositely with the beam. In effect, this composite creates a larger and stronger beam than the steel beam alone. The structural engineer may calculate a transformed section as one step in analyzing the load carrying capability of the composite beam.[2]

Common structural members of composite construction include ① Steel-Concrete Composite Beams

Fig. 5-1 Bank of China Tower in Hong Kong.

(Steel-Concrete Floor Composite Structure); ② Composite Slabs; ③ Steel Reinforced Concrete: Steel Reinforced Beams, Steel Reinforced Columns, Steel Reinforced Walls; ④ Concrete Filled Tube

5.1.1 Composite Steel-Concrete Beam

Composite steel-concrete beam is made up of steel beam and slab by shear connector. In this type of composite beam, concrete and steel beam have joint stress and compatibility of deformation, which makes full use of the compressive strength of concrete and the tensile strength of steel.[3] So it remarkably raises the stiffness and stability of beams.

Composite steel-concrete beams have been developed very fast in recent years, and more and more widely used as the horizontal elements carrying heavy loads in buildings and bridges, obtaining notable economic and social benefits.

Advantages of composite beams:

1. The concrete and steel are utilized effectively.

2. More economical steel section is used in composite construction than conventional non-composite construction for the same span and loading.

3. Depth and weight of steel beam require to be reduced. So, the construction depth also reduces increasing the headroom of the building.

4. Composite beams have higher stiffness, thus they have less deflection than steel beams.

5. Composite beams (Fig. 5-2) can cover for large space without the need of any intermediate columns.

6. Composite construction is faster because of using rolled steel and pre-fabricated components than cast-in-situ concrete.[4]

7. Encased steel beam have higher resistance to fire and corrosion.

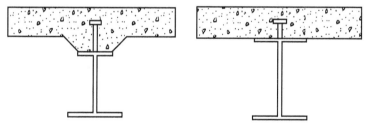

Fig. 5-2 Sections of composite beams.

5.1.2 Composite Slabs

Composite slabs commonly refer to steel deck reinforced composite slabs (Fig. 5-3) which are formed by pouring concrete on the steel plate of different kinds of concave and convex ribs or different kinds grooves. They rely on various forms of concave and convex ribs or grooves and then steel and concrete will be connected together.

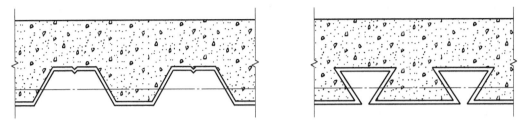

Fig. 5-3 Sections of steel deck reinforced composite slabs.

5.1.3 Steel Reinforced Concrete (SRC)

Steel reinforced concrete (Fig. 5-4) composite structure (short as the SRC composite structure) is made up of profile steel, longitudinal reinforcement, stirrups and concrete. That is concrete structure with profile steel in the core of structure components and with stirrups and appropriate vertical reinforcement out of structure components.[5] It is divided into two types of structure: All structure components are steel reinforced concrete composite structures and part of structure components are steel reinforced concrete composite structures.

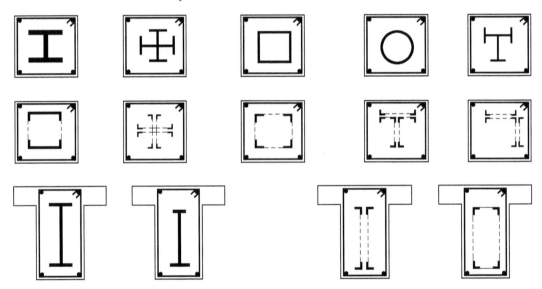

Fig. 5-4 Common sections of SRC.

5.1.4 Concrete Filled Tube (CFT) Columns

Concrete-filled tube columns consist of a steel tube filled with concrete. The concrete core adds stiffness and compressive strength to the tubular column and reduces the potential for inward local buckling.[6] Conversely, the steel tube acts as longitudinal and lateral reinforcement for the concrete core helping it to resist tension, bending moment and shear and providing confinement for the concrete.

Due to the benefit of composite action of the two materials, the CFT columns provide excellent seismic event resistant structural properties such as high strength, high ductility and large energy absorption capacity. Also, circular hollow sections possess many advantages over open sections, including aesthetic appearance and economy in terms of material costs.

Due to the complexity of connections between steel beams and circular hollow sections, their use in structural steelwork is limited. This is because the use of standard bolting is not feasible and costly unpopular welded connections are the normal solutions.[7]

Concrete-filled steel tube (CFT) columns combine the advantages of ductility, generally associated with steel structures, with the stiffness of a concrete structural system.

The advantages of the concrete-filled steel tube column over other composite systems include:

1. The steel tube provides formwork for the concrete.
2. The concrete prolongs local buckling of the steel tube wall.
3. The tube prohibits excessive concrete spalling.
4. Composite columns add significant stiffness to a frame compared to more traditional steel frame construction.

While many advantages exist, the use of concrete filled tubes in building construction has been limited, in part, to a lack of construction experience, a lack of understanding of the design provisions and the complexity of connection detailing. Consequently, a joint is needed that could utilize the favorable strength and stiffness characteristics of the concrete-filled tube column yet be constructible.

5.2 Pre-stressed Concrete Composite Slabs

5.2.1 Introduction

In all kinds of floors, concrete slabs occupy the dominant position. Concrete slabs can be divided into two major categories, the cast-in-place slab and the precast slab. The cast-in-place slab has good structural integrity, however, it needs on-site concrete pouring procedure, large concrete wet construction, instable construction quality, more scaffolding and templates, high costs and long construction period. The precast slab is produced in the factory and has advantages of less concrete wet construction, stable construction quality, less scaffolding and templates, low construction noise, reducible engineering cost and short construction period. But its structural integrity needs to be improved. Therefore, composite slabs which combine the advantages of cast-in-place slabs and precast slabs become popular. Its bottom panel is manufactured in factory and assembled on the spot.[8] Then the concrete is poured in place to form the composite slab. This slab is provided with advantages of good structural integrity, less scaffolding and templates, short construction period, good economic efficiency, so it has promising development prospects.

5.2.2 Overview of Pre-stressed Concrete Composite Plate

The composite slab with pre-stressed bottom panel is the most widely used composite slab. In China, the pre-stressed thin slab was produced up to 1957. In 1970s, the pre-stressed composite beam and slab were produced with cold-drawn low-carbon steel wires. In 1980s, the standard drawing of pre-stressed composite slabs was compiled. Nowadays, the composite slab has been widely used in the housing construction. Integrated prefabricated hyperstatic structures with composite components, formed by pre-stressed composite slabs and pre-stressed hollow composite beams, have been applied in tall buildings.

The bottom panel of composite slabs mostly employs the first-tensioned pre-stressed panel. The bottom panel can also be used as permanent formwork. According to whether or not the braces are set under bottom panels, the composite slab can be sorted into one-stage loading composite slab and two-stage loading composite slab.

The one-stage loading composite slab has slender bottom panel, so it should be supported by temporary braces before the bottom panels are assembled. The post-poured concrete layer and bottom panel begin to work together after post-poured concrete is hardened. The one-stage loading composite slab is similar with the cast-in-place slab in mechanical behavior.

The bottom panel of two-stage composite slab has enough bending stiffness and needs no temporary braces during construction. So construction duration can be shortened and the project cost can be reduced. Different from the cast-in-place slab, the cross-section dimensions and reinforcements of the composite slab are decided by construction stage, not by service stage.[9] Before the post-poured concrete layer is hardened, it bears no loads. In the meanwhile, the bottom panel itself bears its dead weight, post-poured concrete's dead weight and construction loads. After the post-poured concrete layer is hardened, it begins to work together with bottom panel to form composite slab, which bears the dead weight of post-poured concrete layer, floor surface and ceiling and live load during service stage.

The combined interface is the interface between old and new concrete, which determines whether two concrete layers can work together. It has been specially focused on the research and engineering application. Swallowtail grooves were used to increase bond strength between old and new concrete in England. In 1960s, artificial rough surfaces were adopted in the top surface of the bottom panel in the former Soviet Union. In 1970s, shear reinforcements were embedded in the interfaces in the France and West Germany. In 1990s, grooves were set in the top surface of hollow panel to enhance the bond strength by Wang Xizhe. Mechanical behavior of smooth surfaces, indentation surfaces and binding reinforcements were researched by Hou Jianguo.

In addition, edge joints between the precast bottom panels may affect the integrity of composite slabs. Edge joints disposal has always been focused on the research and engineering application.

5.2.3　The Research and Application of Pre-stressed Composite Slabs

To take full advantage of the pre-stressed composite slab and to make up for its shortages, scholars have improved traditional composite slabs in material, bottom panel pattern and edge joint details. To reduce slab weight, Wu Jin et al. studied the lightweight ceramic reinforced concrete composite slab and find out its shear performance.

The bottom panel pattern determines the bonding behavior of the composite slab and the edge joint details influence its crack resistance.[10] The roughness of bottom panel surface can affect the bond force of composite slabs. According to roughness, the bottom panel can be divided into natural rough bottom panel, artificial rough bottom panel and shear-resistant reinforced bottom panel. According to cross section, the bottom panel can be classified into flat bottom panel, precast ribbed bottom panel, hollow bottom panel and sandwich bottom panel.

① Composite slab with flat bottom panel. The bottom panel of the composite slab with flat bottom panel is pretensioned prestressed solid slab, which is the earliest type of composite slab.

Its bending stiffness is inadequacy and needs temporary braces to support during construction. Additionally, the edge joint detail is main factor in crack resistance and integrity for composite slabs.

② Composite slab with precast ribbed panel. The concrete composite slab with precast ribbed panel enhances bending stiffness by introducing ribs on the plat bottom panel. Fewer temporary braces or even no temporary braces are necessary. Therefore, it can facilitate the construction and can reduce construction period and costs.

According to the rib pattern, bottom panels can be divided into one-way rib bottom panels and two-way ribs bottom panels.

③ Composite slab with hollow bottom panel. The composite slab with hollow bottom panel is light enough for large span slabs. Prestressed hollow panel is usually used as bottom panel.

④ Composite slab with sandwich bottom panel. The composite slab with sandwich bottom panel is fabricated by filling with lightweight materials into the composite slab with hollow bottom panel. This lightweight composite slab has functions of sound insulation and thermal preservation.

5.2.4 Conclusion

The composite slab with flat bottom panel has inadequate structural integrity. Its bond performance and crack resistance of edge joint details need to be improved; the composite slab with bar truss reinforced precast concrete bottom panel has enough bond force, but its storage and transportation are inconvenient; the concrete composite slab with precast ribbed panel has good structural integrity and convenient construction procedure; the composite slab with hollow bottom panel and the composite slab with sandwich bottom panel have the function of sound insulation and thermal preservation, however they are inadequate to resist bending moment and have complex construction procedure. It is considered in this paper that among composite slabs mentioned above, the concrete composite slab with precast ribbed panel has more advantages for improvement.

New Words and Expressions

fiberglass   n.   玻璃纤维
stud   n.   螺柱，螺栓
deflection   n.   挠度，偏移
groove   n.   凹槽
stirrup   n.   箍筋

hollow　*adj.*　空心的
formwork　*n.*　模板
scaffolding　*n.*　脚手架
hyperstatic　*adj.*　超静定的
brace　*n.*　支撑
indentation　*n.*　压痕
roughness　*n.*　粗糙度
pretension　预应力
transportation　*n.*　运输
pre-fabricated　*adj.*　预制的
composite steel-concrete structure　钢-混凝土组合结构
convex rib　凸肋
composite slabs　组合楼板
concrete filled tube　钢管混凝土
steel deck reinforced composite slabs　压型混凝土组合板
pre-stressed bottom panel　预应力底板
cold-drawn low-carbon steel　冷拔低碳钢

Notes

1. If the beam is not connected firmly to the slab, then the slab transfers all of its weight to the beam and the slab contributes nothing to the load carrying capability of the beam.

本句难点解析：本句中 firmly 指"坚定的，坚固的"，板和梁之间需要坚固的连接。

本句大意如下：如果梁和楼板之间的连接不够紧密，楼板只是向梁传递了全部自重和上部荷载的力，楼板对梁的承载力提高没有贡献。

2. The structural engineer may calculate a transformed section as one step in analyzing the load carrying capability of the composite beam.

本句难点解析：本句中 structural engineer 是主语，calculate 是谓语。

本句大意如下：结构工程师在计算组合梁承载力时的一个步骤是计算转换（等效）截面。

3. In this type of composite beam, concrete and steel beam have joint stress and compatibility of deformation, which makes full use of the compressive strength of concrete and the tensile strength of steel.

本句难点解析：which 后面是从句，解释组合梁的优势，其可以充分利用混凝土的受压强度和钢材的受拉强度。

本句大意如下：这种复合梁在钢和混凝土的节点上有应力，并且变形协调，可以充分利用混凝土的抗压强度和钢材的抗拉强度。

4. Composite construction is faster because of using rolled steel and pre-fabricated components than cast-in-situ concrete.

Unit 5 Composite Construction

本句难点解析：本句的主干是 Composite construction is faster than cast-in-situ concrete, because of 说明 faster 的原因。

本句大意如下：由于使用轧制钢材和预制构件，因此组合结构施工比现场浇筑的建筑工期短。

5. That is concrete structure with profile steel in the core of structure components and with stirrups and appropriate vertical reinforcement out of structure components.

本句难点解析：and 连接两个句子。这句话在介绍一种新的钢-混凝土组合结构。

本句大意如下：这种混凝土结构在其核心置有型钢，在外部配置箍筋和适当的纵筋。

6. The concrete core adds stiffness and compressive strength to the tubular column and reduces the potential for inward local buckling.

本句难点解析：本句主语是 concrete core，谓语是 adds，混凝土对提高柱子整体的性能有很大作用。

本句大意如下：混凝土核心提高了钢管柱子的刚度和受压强度，减小了向内局部屈曲的趋势。

7. This is because the use of standard bolting is not feasible and costly unpopular welded connections are the normal solution.

本句大意如下：这是因为用标准螺栓连接不可行，一般要用费用较高的不常用的焊接连接。

8. Its bottom panel is manufactured in factory and assembled on the spot.

本句难点解析：联系上下文，本句在说明这种构件可以大大缩短工期的原因。

本句大意如下：它的底板在工厂中制造，在现场装配。

9. Different from the cast-in-place slab, the cross-section dimensions and reinforcements of the composite slab is decided by construction stage, not by service stage.

本句难点解析：本句的主语是 composite slab（组合楼板），谓语是 is。

本句大意如下：和现场浇筑的楼板不同，组合楼板的配筋和截面是由建造阶段决定的，而不是由使用阶段决定的。

10. The bottom panel pattern determines the bonding behavior of the composite slab and the edge joint details influences its crack resistance.

本句难点解析：本句的主语是 bottom panel pattern 和 edge joint details，他们分别影响了组合楼板的不同性能。

本句大意如下：底板的形式决定了组合楼板的粘结性能；边缘节点影响其抗裂性能。

 **Exercises**

Translate the following phrases in to Chinese.

1. Structural point of view
2. bound together strongly
3. load carrying capability
4. dominant position

5. structural integrity

6. two-stage loading composite slab

7. concrete composite slab with flat bottom panel

8. precast ribbed panel

9. edge joint detail

10. complex construction procedure

Translate the following sentences into chinese.

1. Composite construction is a method of construction that is used in a variety of engineering and building applications.

2. In structural engineering, composite construction exists when two different materials are bound together so strongly that they act together as a single unit from a structural point of view.

3. The one-stage loading composite slab has slender bottom panel, so it should be supported by temporary braces before the bottom panels are assembled.

4. The bottom panel of the composite slab with flat bottom panel is pretensioned prestressed solid slab, which is the earliest type of composite slab.

5. It is considered in this paper that among composite slabs mentioned above, the concrete composite slab with precast ribbed panel has more advantages for improvement.

Unit 6

Introduction to Foundation Analysis and Design

6.1 Foundations—Definition and Purpose

All structures designed to be supported by the earth, including buildings, bridges, earth fills, and concrete dams, consist of two parts. These are the superstructure, or upper part, and the substructure element which interfaces the superstructure and supporting ground. In the case of earth fills and dams, there is often not a clear line of demarcation between the superstructure and substructure. The foundation can be defined as the substructure and that adjacent zone of soil and/or rock which will be affected by both the substructure element and its loads.

The foundation engineer is that person who by reason of experience and training can produce solutions for design problems involving this part of the engineered system. In this context, foundation engineering can be defined as the science and art of applying the principles of soil and structural mechanics together with engineering judgment (the "art") to solve the interfacing problem. The foundation engineer is concerned directly with the structural members which affect the transfer of load from the superstructure to the soil such that the resulting soil stability and estimated deformations are tolerable. Since the design geometry and location of the substructure element often have an effect on how the soil responds, the foundation engineer must be reasonably versed in structural design.

A number of practical considerations are a part of the engineering of a foundation:

1. Visual integration of geologic evidence at a site with any field or laboratory test data.[1]
2. Establishing of an adequate field exploration and laboratory testing program.
3. Design of the substructure elements so that they can be built—and as economically as possible.
4. Appreciation of practical construction methods and of likely—to—be—obtained construction tolerances. Stipulation of very close tolerances can have an enormous effect on the foundation costs.

These several items are not directly quantifiable and thus require a considerable application of common sense.

A thorough understanding of the principles of soil mechanics in terms of stability, deformations, and water flow is a necessary ingredient to the successful practice of foundation engineering. Of nearly equal importance is an understanding of the geological processes involved in the formation of soil masses. It is now recognized that both soil stability and deformation are dependent on the stress history of the mass.² It has been common until recently to associate foundation engineering solely with soil mechanics concerns, leaving the interfacing elements to the structural (or other) designer. Current trends are to recognize that foundation engineering is a systems problem and cannot be nicely compartmentalized as some persons would prefer. Readers may determine the validity of this statement as they progress through the text.

The science of soil mechanics and its relationship to geological processes have progressed considerably over the past fifty years. However, because of the natural variability of soil and the resulting problems associated with testing, the design of a foundation still depends to a large degree upon "art", or the application of engineering judgment.³ A subset of this application is the assessment of the tolerable risk associated with the foundation.

The primary focus of this text will be on analysis and design of the interfacing elements for buildings and retaining structures and those soil mechanics principles particularly applicable to these elements. These interfacing elements include both near surface members such as footings and mats and deep elements such as piles and caissons. Retaining structures of concrete (commonly termed retaining walls) and metal (as sheetpiling) are considered in later units. Soil mechanics principles include both stability, including soil water effects, and deformation analyses. Soil stability can often be enhanced by various improvement techniques, the most common being compaction.⁴

■ 6.2 Foundation Classifications

Foundations for structures such as buildings, from the smallest residential to the tallest high—rise, and bridges are for the purpose of transmitting the superstructure loads. These loads come from columns—type members with stress intensities ranging from perhaps 140 MPa for steel to 10 MPa for concrete to the supporting capacity of the soil, which is seldom over 500 kPa but more often on the order of 200 to 250 kPa. The reader can readily note that this interface connects materials whose differences in useful engineering strength can vary by a factor of several hundred. The transmission of these large superstructure loads to the soil may be by use of:

1. Shallow foundations—termed footings, spread footings, or mats. Foundation depth is generally $D \leq B$.

2. Deep foundations—piles or caissons with $D > (4 \sim 5)B$.

Any structure used to retain a soil or similar mass such as grain, coal, or ore in a geometric shape other than that occurring naturally under the influence of gravity is a retaining structure.⁵ Any foundation not classed as shallow, deep, or a retaining structure may be termed a special foundation.

Typical foundation types are:

1. Foundations for buildings (either shallow or deep).

2. Foundations for smokestacks, radio and television towers, bridge piers, industrial equipment, etc. (either shallow or deep).

3. Foundations for port or marine structures (may be shallow or deep and with

基础分类

retaining structures extensively used).

4. Foundations for rotating, reciprocating, and impact machinery, and for turbines, generators, etc. (either shallow or deep and may require vibration control).

5. Foundation elements to support excavation or retain earth masses as for bridge abutments and piers, or retain grain, ore, coal, etc. (retaining walls or sheet-pile structures).

Foundations for buildings are extremely numerous; foundations for the several other types of superstructures are constructed in somewhat lesser numbers.

6.3 Foundation Site and System Economics

A building foundation must be adequate if the structure is to perform satisfactorily and be safe for occupancy. Other foundations must be adequate to perform their intended functions in a satisfactory and safe manner; however, buildings usually have more stringent criteria for safety and performance than other structures-notable exceptions being nuclear-plant facilities, turbines for power generation, and certain types of radio-antenna equipment. Foundations for nuclear plants require extremely rigid design/performance criteria for safety reasons. The other foundations support extremely expensive machinery which is often very sensitive to small soil deformations.

More recently, and after loss of life from several avoidable failures, dam designs where soil is the principal construction material are being more carefully made. One might note that more principles of soil mechanics and geology apply to earth dams than to the majority of foundation engineering problems. In addition to the stringent criteria of the superstructure, instability and water flow through the base soil are serious considerations. A further area of concern is the inevitable deformation of the base soil and subsidence in the superstructure (dam fill material). Careful attention to the occurrence of the latter deformations can allow the designer to avoid a base crack in the dam and the resulting piping failure, or a crest crack and the associated overtopping failure.

Almost any reasonable structure can be built and safely supported if there is unlimited financing. Unfortunately, in the real situation this is seldom, if ever, the case, and the foundation engineer has the dilemma of making a decision under much less than the ideal condition. Also, even though the mistake may be buried, the results from the error are not and can show up relatively soon—and probably before any statute of limitations expires. There are reported cases where the foundation defects (such as cracked walls or broken mechanical fixtures) have shown up years later—also cases where the defects have shown up either during construction of the superstructure or immediately thereafter.

Since the substructure is buried, or is beneath the superstructure, in such a configuration that access will be difficult should foundation inadequacies develop after the superstructure is in place, it is common practice to be conservative. <u>A one or two percent overdesign in these areas produces a larger potential investment return than in the superstructure.</u>[6]

The designer is always faced with the question of what constitutes a safe, economical design while simultaneously contending with the inevitable natural soil het-

基础选址和
经济性

erogeneity at a site. Nowadays that problem may be compounded by land scarcity requiring reclamation of areas which have been used as sanitary landfills, garbage dumps, or even hazardous waste disposal areas. Still another complicating factor is that the act of construction can alter the soil properties considerably from those used in the initial analyses/design of the foundation.[7] These factors result in foundation design becoming so subjective and difficult to quantify that two design firms might come up with completely different designs that would perform equally satisfactory. Cost would likely be the distinguishing feature for the preferred design.

This problem and the widely differing solutions would depend, for example, on the followings:

1. What constitutes satisfactory and tolerable settlement; how much extra could, or should, be spent to reduce estimated settlements from 30 to 15 mm?

2. Has the client been willing to authorize an adequate soil exploration program? What kind of soil variability did the soil borings indicate? Would additional borings actually improve the foundation recommendations?

3. Can the building be supported by the soil using?

a. Spread footings—least cost.

b. Mats—intermediate in cost.

c. Piles or caissons—several times the cost of spread footings.

4. What are the consequences of a foundation failure in terms of public safety? What is the likelihood of a lawsuit if the foundation dose not perform adequately?

5. Is sufficient money available for the foundation? It is not unheard of that the foundation alone would cost so much that the project is not economically feasible. It may be necessary to abandon the site in favor of one where foundation costs are affordable.

6. What is the ability of the local construction force? It is hardly sensible to design an elaborate foundation if no one can build it, or if it is so different in design that the contractor includes a large "uncertain" factor in the bid.

7. What is the engineering ability of the foundation engineer? While this factor is listed last, this is not of least importance in economical design. Obviously, engineers have different levels of capability just as in other professions (lawyers, doctors, professors, etc.) and in the trades, such as carpenters, electricians, and painters.

If the foundation fails because of any cost shaving (in reality implicitly accepting a higher risk), the client tends to quickly lose appreciation for the temporary financial benefit which accrued. At this point, facing heavy damages and/or a lawsuit, the client is probably in the poorest mental state of all the involved parties. Thus, one should always bear in mind that absolute dollar economics may not produce good foundation engineering.

The foundation engineer must look at the entire system: the building purpose, probable service—life loading, type of framing, soil profile, construction methods, and construction costs to arrive at a design that is consistent with the client/owner's needs and dose not excessively degrade the environment. This must be done with a safety factor which produces a tolerable risk level to both the public and the owner.

Considering these several areas of uncertainty, it follows that risk and liability insurance for persons engaged in foundation engineering is very costly. In attempts to reduce these costs as well as produce a design which could be obtained from several engineering firms (i. e. a "consensus" design) there is active discussion (and the practice has already been undertaken in several areas) of having the foundation engineer submit the proposed design to a board of qualified engineers for a "peer review."

6.4 General Requirements of Foundations

A foundation must be capable of satisfying several stability and deformation requirements such as:

1. Depth must be adequate to avoid lateral expulsion of material from beneath the foundation—particularly for footings and mats.

2. <u>Depth must be below the zone of seasonal volume changes caused by freezing, thawing, and plant growth.</u>[8]

3. System must be safe against overturning, rotation, sliding, or soil rupture (shear-strength failure).

4. System must be safe against corrosion or deterioration due to harmful materials present in the soil. This is a particular concern in reclaiming sanitary landfills and sometimes for marine foundations.

5. System should be adequate to sustain some changes in later site or construction geometry, and be easily modified should later changes be major in scope.

6. The foundation should be economical in terms of the method of installation.

7. Total earth movements (generally settlements) and differential movements should be tolerable for both the foundation and superstructure elements.

8. The foundation, and its construction, must meet environmental protection standards.

6.5 Foundation Selection

The principle of foundation selection is listed in Table 6-1.

Table 6-1 Minimum uniformly distributed live loads

Foundation type	Use	Applicable soil conditions
Spread footing, wall footings	Individual columns, walls, bridge piers	Any conditions where bearing capacity is adequate for applied load. May use on single stratum: firm layer over soft layer or soft layer over firm layer. Check immediate, differential, and consolidation settlements
Mat foundation	Same as spread and wall footings. Very heavy column loads. Usually reduces differential settlements and total settlements	Generally, soil bearing value is less than for spread footings over one-half area of building covered by individual footings. Check settlements

Continued

Foundation type	Use	Applicable soil conditions
Pile foundations	In groups (at least 2) to carry heavy column, wall loads; require pile caps	Poor surface and near surface soils. Soils of high bearing capacity 20-50 m below basement or ground surface, but by disturbing load along pile shaft soil strength is adequate. Corrosive soil may require use of timber or concrete pile material.
Bearing	In groups (at least 2) to carry heavy column, wall loads; requires pile cap	Poor surface and near-surface Soils; Soil of high bearing capacity (point bearing on) is 8~50m below ground surface
Caisson (shafts 75 cm or more in diameter) generally bearing or combination of bearing and skin resistance	Larger column loads than for piles but eliminates pile cap by using caissons as column extension	Poor surface and near-surface soils; soil of high bearing capacity (point bearing on) is 8~50m below ground surface
Retaining walls, bridge abutments	Permanent retaining structure	Any type of soil, but a specified zone in back of wall is usually of contracted backfill
Sheet-pile structures	Temporary retaining structures as excavations. Waterfront structures, cofferdams	Any soil; Waterfront structures may require special alloy corrosion protection. Cofferdams require control of fill material

It will be useful, however, at this point to enumerate the several types and their potential application. Where groundwater is present, it is understood that if the depth is below the depth of the footing (or excavation), it will not be a problem. <u>If groundwater is within the construction zone, it must be removed by pumping down the water table, using grout or concrete curtain walls, steel shells, or other means as appropriate.</u>[9]

When groundwater is removed, or when construction is below the water table such that the groundwater may become polluted, approval of appropriate governmental agencies is generally involved to minimize the environmental effects.

6.6 SI and Fps Units

This textbook will use both Foot—pound—second (Fps) and SI units. The SI units will be both those generally accepted and "preferred usage". Problems will be either SI or Fps-there will be no intermixing. In the text either set of units may be used but the alternate system units will not be included in brackets as is commonly done in a number of publications. This method of usage will help the reader in the transition from Fps to SI by requiring concurrent thinking in both sets of units.

Preferred usage will entail a number of mass and pressure units. The reason is that most soil laboratory equipment lasts for years (scales, pressure gages, calipers, etc.) and few laboratories (even in SI countries) have this equipment in true SI units.

In this text, we will define *density* as a mass unit [kg/m^3, kg/cm^3, or g/cm^3 and lb/ft^3 (pcf)]. *Unit weight* is a force unit and in units of pounds or $kips/ft^3$ or $kilonewton/m^3$ (kN/m^3). The SI unit of kilonewton will be used for most if not all soil quantities. This is because the newton is too small and the meganewton is too large (except for steel and concrete stresses).

Using kN will give soil pressures such as intergranular and bearing capacities in kN/m^2 (kilopascal, kPa). In the Fps system, we have correspondingly the pressure unit of $kips/ft^2$ (ksf).

A conversion factor set which is very useful to remember is that to convert mass density in g/cm^3 to kN/m^3 or pcf.

$1 g/cm^3 = 9.807 kN/m^3$ (actually 9.80655 but rounded)

$1 g/cm^3 = 62.4 pcf$

$1 ksf = 47.88 kPa$ (50kPa for general use).

The unit weight of water is 62.4 pcf or $9.807 kN/m^3$, and has a density of $1 g/cm^3$ or $1000 kg/m^3$. Soil will normally vary from

Fps: 90 to 130pcf

SI: 14 to $21 kN/m^3$

Note: If the unit weight γ is reported to 0.1 pcf, then one should report to $0.01 kN/m^3$ for comparable accuracy.

The Fps units of load in tons and pressure in $tons/ft^2$ have been rather widely used in the past both in the field and for certain laboratory tests. This was because the widely used European unit of $1 kg/cm^2$ is very nearly $1 ton/ft^2$ (see Table 6-2) using the 2000lb ton. This text will use only kips, pounds, and kilonewtons for these units, primarily for consistency. The reader should carefully note the source and context when the "ton" unit is used. Sometimes, but not always, it is a metric ton of 1000kg. In some literature, the metric ton is spelled "tone" and represents the long ton of 2204lb.

Table 6-2 Useful SI and metric factors for converting Fps units

To convert from	To	Multiply by
inch	centimeter (cm)	2.54
square inch	square centimeter (cm^2)	6.45160
cubic inch	cubic centimeter (cm^3)	16.38706
in^4	cm^4	41.62314
kilogram force, kgf	newton (N)	9.80665
pound force, lbf	kgf	0.45359
kip	kilonewton (kN)	4.44822
kg/cm^2	kN/m^2 (kPa)	98.0665
kip/in^2 (ksi)	kN/m^2 (kPa)	6894.757
lb/in^2 (psi)	kg/cm^2	0.07031
ton/ft^2	kg/cm^2	0.97653
foot · kip	kN · meter	1.35582

6.7 Computational Accuracy Versus Design Precision

The pocket and desktop calculators, and digital computers, compute with 10 to 14 digits of accuracy. This gives a fictitiously high precision to computed quantities whose input values may have a design precision only within 10 to 30 percent of the numerical value used. The reader should be aware of this actual versus computed precision when checking the example data and output. The author has attempted to maintain a checkable precision by writing the intermediate value (when a pocket calculator was used) to the precision the user should input to check the succeeding steps. If this intermediate value is not used, the computed answer can easily differ in the 0.1 position. The reader should also be aware that typesetting, transcribing, and typing errors inevitably occur, particularly in misplaced parentheses and misleading 3 for 8, etc.

The text user should be able to reproduce most of the digits in the example problems to 0.1 or less unless a typesetting (or other) error has inadvertently occurred. There may be larger discrepancies if the reader uses interpolated data and the author used "exact" data from a computer printout without interpolating.[10] Generally, this will be noted so that the reader is alerted of potential computational discrepancies. The example problems have been included primarily for procedure rather than numerical accuracy, and the user should be aware of this underlying philosophy when studying them.

New Words and Expressions

fill   *n.*   填土，填料，路堤，垫板
demarcation   *n.*   界线分界，边境线
integration   *n.*   整体，结合，综合，一体化
compartmentalize   *v.*   划分，把……分成各自独立的几部分
ore   *n.*   矿物
smokestack   *n.*   烟囱
turbine   *n.*   涡轮机，叶轮机，水（汽）轮机
radio-antenna   *n.*   无线电天线
heterogeneity   *n.*   不均匀性
reclamation   *n.*   回收
lawsuit   *n.*   诉讼
enumerate   *v.*   列举，数
caliper   卡尺，卡钳，测径器，圆规
metric ton   公吨
typesetting   *n.*   排字
interpolate   *v.*   插值，内插，内推

Notes

1. Visual integration of geologic evidence at a site with any field or laboratory test data.

本句难点解析：integration 指 "集成，综合"。联系上文，本句在介绍基础工程的实操需要考虑的因素，其一就是现场直观判断，即本句介绍的内容。

本句大意如下：将现场的地质情况进行可视化整合，包含原位测试或者实验室测得的数据。

2. It is now recognized that both soil stability and deformation are dependent on the stress histo-

Unit 6　Introduction to Foundation Analysis and Design

ry of the mass.

本句难点解析：stability 指"稳定性"。解释了土的性质与土的应力历史有关。

本句大意如下：现在认识到土体的稳定性和变形与土的应力历史有关。

3. However, because of the natural variability of soil and the resulting problems associated with testing, the design of a foundation still depends to a large degree upon "art", or the application of engineering judgment.

本句难点解析：这句话前面用 because 解释了影响基础的设计的难点。

本句大意如下：然而，由于土本身的特性多种多样，由此导致试验中的各种问题，基础的设计很大程度上依靠"艺术"，或者依靠工程师的判断。

4. Soil stability can often be enhanced by various improvement techniques, the most common being compaction.

本句难点解析：本句主语是 soil stability，be enhanced by 后面是提高土体稳定性的方法。

本句大意如下：土体的稳定性可以有很多办法提高，最常用的是压实法。

5. Any structure used to retain a soil or similar mass such as grain, coal, or ore in a geometric shape other than that occurring naturally under the influence of gravity is a retaining structure.

本句难点解析：句子的主体是 Any structure is a retaining structure，任何维持 grain, coal, or ore（颗粒，煤炭，矿物）等几何形状的结构等内容都是定语，说明 any structure 的特点。

本句大意如下：挡土结构是用于维持土或者类似质量块（例如颗粒、煤炭或者矿物）保持一个几何形状的结构，这种几何形状区别于在重力作用下形成的天然的形状。

6. A one or two percent overdesign in these areas produces a larger potential investment return than in the superstructure.

本句难点解析：A one or two percent overdesign 是主语，produces 是谓语。

本句大意如下：比起上部结构保险（安全）设计，基础的一到两个百分比的保险（安全）设计会带来更大的潜在投资回报。

7. Still another complicating factor is that the act of construction can alter the soil properties considerably from those used in the initial analyses/design of the foundation.

本句难点解析：主语是 another complicating factor，从句中解释了基础建设复杂的本质原因是土的性质发生改变。

本句大意如下：还有其他复杂的原因是，施工会显著改变土体性质，使之与基础最初设计/分析的时候不同。

8. Depth must be below the zone of seasonal volume changes caused by freezing, thawing, and plant growth.

本句难点解析：seasonal volume change 指由于季节变换原因引起的土体体积的改变。

本句大意如下：基础的埋深要低于季节体积变化区域，土体季节体积变化主要由于冰冻、融化和植物生长引起。

9. If groundwater is within the construction zone, it must be removed by pumping down the water table, using grout or concrete curtain walls, steel shells, or other means as appropriate.

本句难点解析：using 后面接了三种合适的方法来解决施工区域存在地下水的问题：using grout or concrete curtain walls, steel shells（使用灌浆，混凝土幕墙，钢壳）等方式本句大意如下：如果在施工区域有地下水，必须通过有效的方法先把水排到允许水位线以下，可以采用灌浆、混凝土幕墙、钢壳等方式。

10. There may be larger discrepancies if the reader uses interpolated data and the author used "exact" data from a computer printout without interpolating.

本句难点解析：interpolated 在这里解释为"插值"，large discrepancies 在这里是较大的误差。

本句大意如下：运用插值得到的数据与计算机输出的没有经过插值的准确数据相比可能会有很大的偏差。

Translate the following phrases into Chinese.

1. adjacent zone of soil
2. laboratory testing program
3. soil masses
4. soil mechanics
5. stringent criteria of the superstructure
6. sanitary landfills
7. safe against overturning and sliding
8. curtain wall
9. checkable precision
10. underlying philosophy

Translate the following sentences in to Chinese.

1. The foundation can be defined as the substructure and that adjacent zone of soil and/or rock which will be affected by both the substructure element and its loads.

2. The reader can readily note that this interface connects materials whose differences in useful engineering strength can vary by a factor of several hundred.

3. In addition to the stringent criteria of the superstructure, instability and water flow through the base soil are serious considerations.

4. This method of usage will help the reader in the transition from Fps to SI by requiring concurrent thinking in both sets of units.

5. Generally, this will be noted so that the reader is alerted of potential computational discrepancies.

Part 2

Introduction to Other Branches of Civil Engineering

Unit 7

Introduction to Bridge Engineering

7.1 Reinforced Concrete Girder Bridges

The raw materials of concrete, consisting of water, fine aggregate, coarse aggregate, and cement, can be found in most areas of the world and can be mixed to form a variety of structural shapes. The great availability and flexibility of concrete material and reinforcing bars have made the reinforced concrete bridge a very competitive alternative. Reinforced concrete bridges may consist of precast concrete elements, which are fabricated at a production plant and then transported for erection at the job site, or cast-in-place concrete, which is formed and cast directly in its setting location.[1] Cast-in-place concrete structures are often constructed monolithically and continuously. They usually provide a relatively low maintenance cost and better earthquake-resistance performance. Cast-in-place concrete structures, however, may not be a good choice when the project is on a fast-track construction schedule or when the available falsework opening clearance is limited.[2] In this unit, various structural types and design considerations for conventional cast-in-place, reinforced concrete highway girder bridge are discussed.

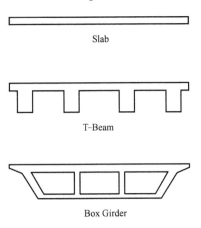

Fig. 7-1 Typical reinforced concrete sections in bridge superstructures.

Reinforced concrete sections, used in the bridge superstructures, usually consist of slabs, T-beams (deck girders), and box girders (Fig. 7-1). Safety, cost-effectiveness, and aesthetics are generally the controlling factors in the selection of the proper type of bridges. Occasionally, the selection is complicated by other considerations such as the deflection limit, life-cycle cost, traffic maintenance during construction stages, construction scheduling and worker safety, feasibility of falsework layout, passage of flood debris, seismicity at the site, suitability for future widening, and commit-menu made to officials and individuals of the community.[3] In some cases, a prestressed concrete or steel bridge may be a better choice.

1. Slab Bridges

Longitudinally reinforced slab bridges have the simplest superstructure configuration and the neatest appearance. They generally require more reinforcing steel and structural concrete than do girder-type bridges of the same span. However, the design details and formworks are easier and less expensive. It has been found economical for simply supported spans up to 9 m and for continuous spans up to 12 m.

2. T-Beam Bridges

The T-beam construction consists of a transversely reinforced slab deck which spans across to the longitudinal support girders. These require a more-complicated formwork, particularly for skewed bridges, compared to the other superstructure forms. T-beam bridges are generally more economical for spans of 12 to 18 m. The girder stem thickness usually varies from 35 to 55cm and is controlled by the required horizontal spacing of the positive moment reinforcement. Optimum lateral spacing of longitudinal girders is typically between 1.8 and 3.0m for a minimum cost of formwork and structural materials.[4] However, where vertical supports for the formwork are difficult and expensive, girder spacing can be increased accordingly.

3. Box-Girder Bridges

Box-girder bridges contain top deck, vertical web, and bottom slab and are often used for spans of 15 to 36 m with girders spaced at 1.5 times the structure depth. Beyond this range, it is probably more economical to consider a different type of bridge, such as post-tensioned box girder or steel-girder superstructure. This is because of the massive increase in volume and materials. They can be viewed as T-beam structures for both positive and negative moments. The high torsional strength of the box girder makes it particularly suitable for sharp curve alignment, skewed piers and abutments, superelevation, and transitions such as interchange ramp structures.[5]

7.2 Arch Bridges

The origins of the use of arches as a structural form in buildings can be traced back to antiquity (Van Beek, 1987). In trying to arrive at a suitable definition for an arch we may look no further than Hooke's anagram of 1675 which stated "Ut pendet continuum flexile, sic stabat continuum rigidum inversum" — "as hangs the flexible line, so but inverted will stand the rigid arch". This suggests that any given loading to a flexible cable if frozen and inverted will provide a purely compressive structure in equilibrium with the applied load. Clearly, any slight variation in the loading will result in a moment being induced in the arch. It is arriving at appropriate proportions of arch thickness to accommodate the range of eccentricities of the thrust line that is the challenge to the bridge engineer.

Even in the Middle Ages, it was appreciated that masonry arches behaved essentially as gravity structures, for which geometry and proportion dictated aesthetic appeal and stability.[6] Compressive strength could be relied upon whilst tensile strength could not. Based upon experience, many empirical relationships between the span and arch thickness were developed and applied successfully to produce many elegant structures throughout Europe. The expansion of the railway and canal systems led to an explosion of bridge building. Brickwork arches became increasingly popular. With the construction of the Coal-brookdale Bridge (1780) a new era of arch bridge construction began. By the end of the nineteenth century cast iron, wrought iron and finally steel became increasingly popular; only to be challenged by ferrocement (reinforced concrete) at the turn of the century.

During the nineteenth century analytical technique developed apace. In particular, Castigliano (1879) developed strain energy theorems which could be applied to arches provided that they remained elastic. This condition could be satisfied provided the line of thrust lay within the middle third, thus ensuring that no tensile stresses were induced. The requirement to avoid tensile stresses only applied to masonry and cast iron; it did not apply to steel or reinforced concrete (or timber for that matter) as these materials were capable of resisting tensile stresses. Twentieth century arch bridges have become increasingly sophisticated structures combining modern materials to create exciting functional urban sculptures.

7.3 Steel Bridges

Structural steel is an extremely versatile material eminently suited for the construction of all forms of bridges. The material, which has a high strength-to-weight ratio, can be used to bridge a range of spans from short through to very long (15~1500m), supporting the imposed loads with the minimum of dead weight. Steel bridges normally result in light superstructures which in turn lead to smaller, economical foundations. They are normally prefabricated in sections in a factory environment under strict quality control, transported to site in manageable units and bolted together in situ to form the complete bridge structure. Using this construction method the erection of a steel bridge is usually rapid, resulting in minimal disruption to traffic; a very important factor if traffic delays, be it road or rail, are properly assessed in the construction project. Rolled steel sections, the largest manufactured in the UK being a 914×419×388 kg/m universal beam, are economical for short-span highway bridges and, when designed to act compositely with the concrete deck, are capable of spanning 25~30m. For spans in the range of about 25~100m, plate girders, again acting compositely with the deck, provide an economical solution. In order to optimise the concrete deck, which has to distribute wheel loads transversely across the bridge, it is usual to arrange for a plate girder spacing of around 3m. For longer spans exceeding 100m, box girders are the favoured choices. Although box girders have a higher fabrication cost than plate girders, box girders have substantially greater torsional stiffness and, if carefully profiled, good aerodynamic stability.[7] For very long spans in excess of 250m, stiffened steel box girders with an integral orthotropically stiffened steel top plate, forming the primary support for the running surface, provide a very economical lightweight solution. Fig. 7-2 gives cross-sections taken through typical bridge structures using hot-rolled, plate girder and box girder sections.

7.4 Truss Bridges

Lattice truss structures have been used very successfully for both railway and highway bridges throughout the last 150 years. There are three main truss configurations in use today, namely the

Warren truss, the Modified Warren truss and the Pratt truss, all of which can be used as an underslung truss, a semi-through truss, or a through truss bridge. Fig. 7-3 gives details of the three truss types together with sections showing the differences between an underslung, semi-through and through truss.

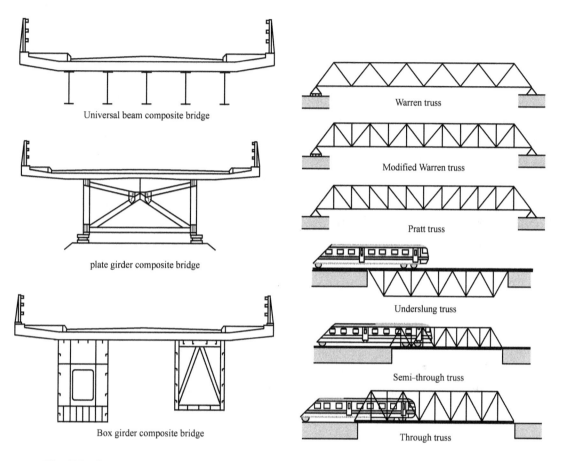

Fig. 7-2 Cross-sections through typical composite bridges.

Fig. 7-3 Typical truss bridges.

7.5 Plate and Box Girder Bridges

As discussed previously in this unit, one of the most common forms of steel (or composite) bridge, the plate girder, is comprised of steel plate elements welded together, often of relatively slender construction. These elements are found in the webs and flanges of plate girders and also in the stiffeners, although the latter are normally made from hot-rolled sections. The box girder, somewhat less common, is found in longer-span bridges either as a composite construction or, for very long spans, as an all-steel structure with stiffened steel decks. Such long-span girders may have additional support provided by cables such as found in cable-stayed or suspension bridges. Very long-span bridges can have extremely complex cross-sections of aerodynamic shape with complex stiffening arrangements. Because of the slender nature of the individual plate components, they are prone to local buckling. In addition, other buckling modes might occur such as lateral torsional buckling of a plate girder between points of lateral flange restraint. This is discussed later in this unit. In order to design the plated elements of a plate or box girder it is desirable to have a degree of understanding of the behaviour of plates and the functions and effects of stiffening under various types of loading.

7.6 Cable Stayed Bridges

The use of inclined stays as a tension support to a bridge deck was a well-known concept in the nineteenth century and there are many examples, particularly using the inclined stay as added stiffness to the primary draped cables of the suspension bridge.[8] Unfortunately, at this time, the concept was not well understood. As it was not possible to tension the stays they would become slack under various load conditions. The structures often had inadequate resistance to wind-induced oscillations. There were several notable collapses of such bridges, for example the bridge over the Tweed River at Dryburgh (Drewry, 1832), built in 1817, and collapsed in 1818 during a gale only six months after construction was completed. As a result the use of the stay concept was abandoned in England. Nevertheless, these ideas were adapted and improved by the American bridge engineer Roebling who used cable stays in conjunction with the draped suspension cable for the design of his bridges. The best known of Roebling's bridges is the Brooklyn Bridge, completed in 1883. The modern concept of the cable-stayed bridge was first proposed in postwar Germany, in the early 1950s, for the reconstruction of a number of bridges over the River Rhine. These bridges proved more economic, for moderate spans, than either the suspension or arch bridge forms. It proved very difficult and expensive in the prevailing soil conditions of an alluvial floodplain to provide the gravity anchorages required for the cables of suspension bridges.[9] Similarly for the arch structure, whether designed with the arch thrust carried at foundation level or carried as a tied arch, substantial foundations were required to carry these large heavy spans. By comparison the cable-stayedalternatives had light decks and the tensile cable forces were part of a closed force system which balanced these forces with the compression within the deck and pylon. Thus expensive external gravity anchorages were not required.

The construction of the modern multi-stay cable-stayed bridge can be seen as an extension, for larger spans, of the prestressed concrete, balanced cantilever form of construction. The tension cables in the cable-stayed bridge are located outside the deck section, and the girder is no longer required to be of variable depth. However, the principle of the balanced cantilever modular erection sequence, where each deck unit is a constant length and erected with the supporting stays in each erection cycle, is retained.[10] The first modern cable-stayed bridge was the Stromsund Bridge (Wenk, 1954) in Sweden constructed by the firm Demag, with the assistance of the German engineer Dischinger, in 1955. At the same time Leonhardt designed the Theodor Heuss Bridge (Beyer and Tussing, 1955) across the Rhine at Dusseldorf but this bridge, also known as the North Bridge, was not constructed until 1958. The first modern cable-stayed bridge constructed in the United Kingdom was the George Street Bridge over the Usk River (Brown, 1966) at Newport, South Wales which was constructed in 1964. These structures were designed with twin vertical stay planes. The first structure with twin inclined planes connected from the edge of the deck to an A-frame pylon was the Severins Bridge (Fischer, 1960) crossing the River Rhine at Cologne, Germany. This bridge was also the first bridge designed as an asymmetrical two-span structure. The economic advantages described above are valid to this day and have established the cable-stayed bridge in its unique position as the preferred bridge concept for major crossings within a wide range of spans. The longest-span cable-stayed bridge so far completed is the Tatara Bridge in Japan with a main span of 890m. At the time of writing (2008) several other bridges are planned, or are in construction, with main spans in excess of 1000m, notably the Sutong Bridge (1088m) and the Stonecutters Bridge (1018m), which are both in China.

■ 7.7 Suspension Bridges

The scope of this unit is restricted to consideration of the classical three-span suspension bridge configuration (Fig. 7-4), with a stiffened load-carrying deck structure supported by earth-anchored cables. The bridge may have side spans of differing lengths and, depending on the site topography, the bridge deck may be suspended either in all three spans or in the main span only, when the side span cables act simply as back stays to the towers. Bridges with unusual span or cable configurations, including bridges with multiple main spans, mono-cable bridges, self-anchored structures, and hybrid part suspension/part cable-stayed structures are not considered. Even with the above limitations, it is not possible in a relatively short chapter to consider in detail many important aspects of suspension bridge design—in particular the analysis of cables and the aerodynamic design requirements. The principal structural elements of the classical three-span suspension bridge are as follows:

Two or more main cables formed from high-strength steel wires, with a strength-to-weight ratio of around three times that of weldable structural steels, and which support the traffic-carrying deck and transfer its loading by direct tension forces to the supporting towers and anchorages.

A deck structure together with a longitudinal stiffening system to distribute concentrated traffic loadings on the deck, and control local deflections. The stiffening system also provides the necessary torsional stiffness to prevent aerodynamic instability of the deck and may be either separate truss or plate stiffening girders combined with lateral bracing systems, or alternatively be integrated with the deck structure in the form of a shallow box girder with a low drag shape to minimise wind loading. As the deck dead load is entirely supported by the cable, towers and anchorages, an economical overall design requires the lightest practicable deck structure which is supported from the main cables by hangers of high-strength wire ropes or strands that are spaced at regular intervals throughout the spans.

Towers to support the main cables at a level determined by the main span cable sag, combined with the required clearance above the waterway or other obstacles being crossed.

Anchorages to secure the ends of the main cables against movement. These structures must resist large horizontal forces and, in areas where ground conditions are poor their cost may be very high, to some extent off setting the economy of the main load-carrying structure.

With cables constructed from very high-strength steel loaded in direct tension as their primary load-carrying members, suspension bridges are ideally suited to longer spans, and this is therefore the primary application for this type of structure. Although cable-stayed structures have made considerable inroads into the span range previously considered to be the domain of suspension bridges, these remain the unchallenged choice for spans over 1200m. When well designed and proportioned, suspension bridges are the most beautiful bridges, as the simplicity of the structural arrangement, the natural curve of the main cables, the slender suspended deck and towers, produce an aesthetically attractive structure. This natural grace can also make suspension bridges a suitable choice for relatively short-span foot bridges in situations where an attractive appearance is an important consideration.

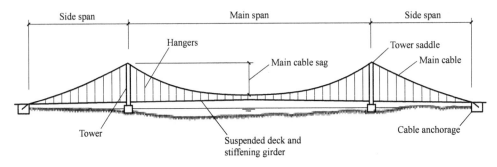

Fig. 7-4 Three-span bridge diagram.

New Words and Expressions

longitudinal *adj.* 经度的纵向的；纵的；纵观的
post-tensioned *adj.* 后张拉的
pier *n.* 码头，防波堤；桥墩
abutment *n.* 邻接，桥墩，桥基；扶垛；对接；接界
ramp *n.* 土堤邻坡；斜道；坡道
precast *adj.* 预浇铸的，预制的；*v.* 预制；预浇制
fabricate *v.* 制造
seismicity *n.* 地震活动；受震强度
superstructure *n.* 上部结构，上层建筑；（建筑物、船等的）上面部分
configuration *n.* 布局，构造；配置
coarse *adj.* 粗鄙的
slack *adj.* 松（弛）的
empirical *adj.* 凭经验的；经验主义的；以观察或试验为依据的
ferrocement *n.* 铁矿渣水泥
masonry *n.* 砖石建筑，砌体结构

Unit 7　Introduction to Bridge Engineering

timber　　n.　木材；（用于建筑或制作物品的）树木；用材林，林场；素质

1. Reinforced concrete bridges may consist of precast concrete elements, which are fabricated at a production plant and then transported for erection at the job site, or cast-in-place concrete, which is formed and cast directly in its setting location.

本句难点解析：这句话的主语是 bridges，谓语是 consist of，宾语是 precast concrete elements or cast-in-place concrete，宾语之后跟 which 引导的定语从句补充说明宾语混凝土构件的预制流程，第二个 which 引导的定语从句用来修饰 cast-in-place concrete。

本句大意如下：钢筋混凝土桥由预制混凝土构件或现浇混凝土组成，预制构件在生产车间中制造后被运送至工地进行安装，现浇混凝土在现场设定的位置支模并直接浇筑成形。

2. Cast-in-place concrete structures, however, may not be a good choice when the project is on a fast-track construction schedule or when the available falsework opening clearance is limited. 本句难点解析：这句话的主语是 cast-in-place concrete structures，谓语是 may not be，宾语是 choice，when 引导的条件状语从句说明了在什么条件下现浇混凝土结构不是一个好的选择。

本句大意如下：然而，当项目施工工期较短或可用脚手架的开口间隙有限时，现浇混凝土结构可能不是一个好的选择。

3. Occasionally, the selection is complicated by other considerations such as the deflection limit, life-cycle cost, traffic maintenance during construction stages, construction scheduling and worker safety, feasibility of falsework layout, passage of flood debris, seismicity at the site, suitability for future widening, and commit-menu made to officials and individuals of the community.

本句难点解析：这句话的主语是 selection，谓语是 is，宾语是 considerations，宾语之后 such as 列举了让选择变复杂的其他因素。

本句大意如下：有时，桥梁类型的选择会因为考虑其他因素而变得复杂，这些因素包括变形极限、全寿命成本、施工阶段的交通维护、施工进度和工人安全、脚手架布置的可行性、泥石流通道、场地的地震活动、未来扩建的适宜性以及对社区官员和个人的承诺内容等。

4. Optimum lateral spacing of longitudinal girders is typically between 1.8 and 3.0m for a minimum cost of formwork and structural materials.

本句难点解析：这句话的主语是 spacing，谓语是 is，宾语是 1.8 and 3.0m。

本句大意如下：纵向主梁的最佳侧向间距通常在 1.8~3 米之间，以获得模板和结构材料的最小成本。

5. The high torsional strength of the box girder makes it particularly suitable for sharp curve alignment, skewed piers and abutments, super elevation, and transitions such as interchange ramp structures.

本句难点解析：这句话的主语是 strength，谓语是 makes，宾语是 sharp curve alignment, skewed piers and abutments, super elevation, and transitions。

本句大意如下：箱型主梁的抗扭强度高，特别适合于急弯校直、偏斜墩、超高程和向立

体坡道结构的过渡段。

6. Even in the Middle Ages, it was appreciated that masonry arches behaved essentially as gravity structures, for which geometry and proportion dictated aesthetic appeal and stability. 本句难点解析：这句话是主语从句，形式主语是 it，实际主语为 that 引导的从句，for which 引导定语从句进一步说明实际主语 masonry arches。本句大意如下：甚至在中世纪，人们就意识到砌体拱本质上是重力结构，这种结构的几何和比例决定了美学魅力和稳定性。

7. Although box girders have a higher fabrication cost than plate girders, box girders have substantially greater torsional stiffness and, if carefully profiled, good aerodynamic stability. 本句难点解析：Although 引导条件状语从句，这句话的主语是 box girders，谓语是 have，if 引导条件状语从句。

本句大意如下：虽然箱型主梁的制造成本比板梁高，但箱梁有相当大的扭转刚度，且如果仔细设计断面，其可以有很好的空气动力稳定性。

8. The use of inclined stays as a tension support to a bridge deck was a well-known concept in the nineteenth century and there are many examples, particularly using the inclined stay as added stiffness to the primary draped cables of the suspension bridge.

本句难点解析：and 作为连词并列两个句子。第一句话的主语是 use，谓语是 was，宾语是 concept；第二句话的主语是 there，谓语是 are，宾语是 examples，particularly using 引导的部分修饰宾语说明这些例子的主要特点。

本句大意如下：斜拉索作为桥面板的受拉支撑的使用在 19 世纪已是一个众所周知的概念并且已有许多实例，特别是利用斜拉索来增加吊桥主要悬索刚度的案例。

9. It proved very difficult and expensive in the prevailing soil conditions of an alluvial floodplain to provide the gravity anchorages required for the cables of suspension bridges.

本句难点解析：这句话主语是 it，谓语是 proved，宾语是 difficult and expensive，to provide 和宾语形成搭配具体说明什么是困难和昂贵的，in 引导的部分说明了这个结论是在某种特殊的土壤条件下。

本句大意如下：事实证明，在冲积平原的土体条件下，为悬索桥的悬索提供重力锚固是非常困难和昂贵的。

10. However, the principle of the balanced cantilever modular erection sequence, where each deck unit is a constant length and erected with the supporting stays in each erection cycle, is retained.

本句难点解析：这句话的主语是 principle，谓语是 is，宾语是 retained，where 引导定语从句修饰 modular erection sequence。

本句大意如下：然而，平衡悬臂模块化装配顺序的原理得以保留，包括每个板单元拥有恒定的长度并且在每个装配周期中与支撑悬索一同装配。

Translate the following phrases into Chinese.

1. coarse aggregate

2. constructed monolithically and continuously

3. earthquake-resistance performance

4. elegant structure

5. long spans

6. semi-through truss

7. wind-induced oscillations

8. suspension bridge

9. mono-cable bridge

10. short-span footbridge

Translate the following sentences into Chinese.

1. The raw materials of concrete, consisting of water, fine aggregate, coarse aggregate, and cement, can be found in most areas of the world and can be mixed to form variety of structural shapes.

2. Beyond this range, it is probably more economical to consider a different type of bridge, such as post-tensioned box girder or steel girder superstructure.

3. Based upon experience, many empirical relationships between the span and arch thickness were developed and applied successfully to produce many elegant structures throughout Europe.

4. Because of the slender nature of the individual plate components, they are prone to local buckling.

5. The construction of the modern multi-stay cable-stayed bridge can be seen as an extension, for larger spans, of the prestressed concrete, balanced cantilever form of construction.

Unit 8

Introduction to Underground Engineering

8.1 The Future of Underground Infrastructure in Holland

8.1.1 Introduction

With several major infrastructure projects planned or in progress, there has been a progressive increase in interest in The Netherlands for underground alternatives for infrastructure facilities. Planners and the construction industry are preparing to apply competitive under ground solutions to all types of infrastructure problems, including application of subsurface tunnelling by means of shield-supported tunnelling installations.[1]

This paper summarises (1) the determining issues for application of underground solutions; and (2) the expectations regarding development of the construction methods for underground solution (especially for the "bottlenecks") of infrastructure projects. The discussion mainly relates to road and railroad infrastructure.

8.1.2 Driving Forces

Driving forces for underground infrastructure (the "positive drivers") are:
The density of built-up areas.
The urban environment.
The limitations on extension of the heavily loaded existing infrastructure facilities.
The possible complexity of realizing infrastructural works.
In an increasing number of cases, infrastructure projects would require the application of a tunnel boring method.

In addition to the above-listed traditional types of positive drivers, there are new ones, which influence the choice in favor of underground alternatives. These include a broad range of environment-related issues and longer-term physical planning strategic aspects.

"Negative drivers", which reduce or even eliminate the attractiveness of underground solutions are: risk; time; and costs—all of which are elements related to construction.

Mastering risk at competitive costs and within competitive time limits is the first criterion for accepting a subsurface option. Subsequently, cost and time are criteria for the selecting subsurface options, with cost as the leading indicator.

There is no doubt that current cost indications represent the major limiting factor for going under-ground, since the cheaper alternatives have not (yet) been sufficiently proven to be considered offering real competition to other types of alternatives.

Promoting underground solution unreasonably or unrealistically is not the way to achieve a fair competition between subsurface and other types of solutions for the realization of infrastructural projects. The best "ambassador" for going underground is the completion of underground infrastructure projects, whatever their nature and however small, within contract conditions. This also provides the best basis for further optimization that will lead to a truly competitive technology for underground solutions.

8.1.3 Challenges

The Netherlands has broad experience with the construction of "in-soil" structures (i. e., partly or entirely below surface without soil cover), "mounded" structures (i. e., partly below surface, covered by earth) and "shallow depth" structures (i. e., below the surface, built by means of a cut-and-fill method). The challenge of the application of this type of structures for underground infrastructure projects on a larger scale is optimization of the method versus functional requirements and cost level, taking into account longer-term (lifetime duration) behavior.

Innovative application of the methods referred to above and/or material usage can no doubt provide cost-effective and in all respects competitive solutions. However, the methods proposed must be reliably controllable in all respects, including longer-term behavior and behavior under extreme conditions. The consequences of discovering shortcomings in a (too) late stage may turn out to be very disadvantageous, at least for the project and/or the future owner/operator.

As already stated, infrastructure projects increasingly require the application of a tunnel boring method, the use of which has been very limited in the Netherlands. It can even be stated that experience with application of a tunnel boring method under conditions similar to the western part of the Netherlands is very scarce, if indeed it exists at all, anywhere in the world.

The unique elements of the Dutch conditions for tunnel boring methods are:

The complexity of the subsoil conditions (i. e., softness, heterogeneity and extreme sensitivity to stress/strain/volumetric changes); the uncertainties with regard to foundations of existing structures (which could be unpredictably overloaded due to negative skin friction and/or differential settlement effects with possible major variations in the load state of neighborliness foundation elements) and the subsequent uncertainties in the structure/foundation interaction; the uncertainties associated with the status of buried utility networks (due to possible major differences in deformation as a consequence of easy occurrence of variation in soil settlements within short distances); and the potential risk of liquefaction occurring under existing roads or railroads.

Challenges with respect to application of tunnel boring methods are of a different order of magnitude than those related to "in-ground", "mounded" or "shallow depth" structures. The problems of predicting the progress and determining the costs could easily be outclassed by the problems of predicting the possible consequences of tunnel boring application. The challenge is that unacceptable consequences must not occur while risk must be taken-up to the limit of acceptability—to permit cost-competitiveness. To accomplish this requires expertise and experience beyond the limits of tunnelling method application, and optimized application of management, quality control and monitoring systems.

8.1.4 Structured Development

In order to enhance the development of underground construction methods to a competitive level versus the other options within the shortest possible period, a development programme has been started with substantial government support, coordinated by the "Centrum Ondergronds Bouwen" (COB, Centre for Underground Space Technology). The Centre is to commission, initiate and coordinate practically all relevant research and development activities on tunnelling under Dutch conditions. Participants in the COB programme are government parties, contractors, consultants, research institutes, universities and (future) owners/operators of infrastructure facilities.

The COB programme provides an excellent opportunity for developing underground space technology to a competitive level under Dutch conditions and within the shortest possible period. The challenge in this respect is to meet expectations and develop, within the proposed time frame, a solid and overall basis—rather than simply sorting out some complex details—for further development of working methods for and widespread application of underground space technology under the Dutch conditions.

8.1.5 DTBM—an Expectation

Based on the expertise of Dutch specialists in geotechnical and structural engineering, the high technical standards and creativity of Dutch contractors, the COB—guided basic development programme, the thoroughness and inventiveness evident during construction of other types of major (first-of-their-kind) civil engineering projects in the Netherlands, and the problems increasingly requiring underground solutions, it maybe expected that various tunnelling methods will be competitively applied under Dutch conditions within a few years.

On the same basis, it is also expected that a Dutch Tunnel Boring Method (DTBM) for large-diameter tunnels will be developed, in all respects fit for realising infrastructure works under the specific Dutch ground conditions. Essential elements of the DTBM could include the followings:

1) An extensive and detailed survey of the subsoil carried out from the tunnelling machine. Such a survey, in itself a major innovative element, will focus on the soil in front of the tunnelling machine. A typical dimension of the soil-plug to be investigated is 3 to 5 D in length and 2 to 3 D in diameter (D = diameter of the proposed tunnel).

2) The possibility of carrying out solidification of the subsoil, if and when necessary, from the tunnelling machine. The subsoil surrounding the proposed tunnel (typically 1 to 2.5 m thick) could be subject to solidification (by means of freezing, cement injection, chemical injection, etc.), most

likely in combination with solidifying (part of) the soil to be removed.

3) Tunnel construction with a tunnel boring machine with high flexibility in soil removal, liner implementation and the additional tasks (soil investigation, soil solidification, etc.). Operation of the machine within its flexibility limits will be dictated by the subsoil conditions and the risks related to the surroundings. Such a machine could work fast and inexpensively under favorable or low-risk conditions for perhaps the largest part of the project, and adapt the working method when conditions deteriorate.

4) A "control unit" which would be responsible for processing the data from the soil survey carried out from the tunnelling machine, cross-checking these data with the previously collected data/information/predictions, and determining the working method for the next part (perhaps 25 m) of the track.

8.1.6 Competitiveness

A fair comparison between an underground solution and other types of alternatives can only be made on the basis of an integrated assessment of all relevant aspects. However, it should be noted that without real competitiveness, the chances of underground alternatives being selected are very remote. In this respect, the following comments should be taken into consideration:

1) Competitiveness requires limiting risk to a controllable level. At the same time, making optimal use of the controllable risk—and this requires considerable expertise—is an essential part of competitive cost formulations.

2) A competitive technology requires optimal adaptation to circumstances within the limits, of controllable risks. A working method based on average conditions is too expensive.

3) Optimisation of the cost level will require lump-sum turnkey contracts, with a clear agreement before contract award regarding uncertainties (which will be numerous when applying underground technology, especially during realisation of the first such projects), risk control, quality control and performance monitoring.[2] Cost optimisation on the Client side should be based on a Functional Value Analysis.

4) Performance monitoring and quality control, to be carried out by or on behalf of the Client, should be well developed and sufficiently detailed, based on the bidding documents. To ensure effective control of projects under discussion requires a balance of expertise between the Contractor and the Client, both possibly supported by advisers.

8.1.7 Conclusions

In spite of the extremely unfavourable subsoil and related conditions, there are promising opportunities to develop competitive methods for underground infrastructure in the Netherlands at this time.

The challenge to the engineering community and contractors in the Netherlands is to use these opportunities in such a way that Dutch industry can take a leading position in soft soil tunnelling within a time frame of 10 years.[3]

8.2　Seismic Design and Analysis of Underground Structure

Underground structures have features that make their seismic behavior distinct from most surface structures, most notably (1) their complete enclosure in soil or rock, and (2) their significant length (i.e. tunnels). The design of underground facilities to withstand seismic loading thus, has aspects that are very different from the seismic design of surface structures.

This report focuses on relatively large underground facilities commonly used in urban areas. This includes large-diameter tunnels, cut-and-cover structures and portal structures (Fig. 8-1). This report does not discuss pipelines or sewer lines, nor does it specifically discuss issues related to deep chambers such as hydropower plants, nuclear waste repositories, mine chambers, and protective structures, though many of the design methods and analyses described are applicable to the design of these deep chambers.[4]

Large-diameter tunnels are linear underground structures in which the length is much larger than the cross-sectional dimension. These structures can be grouped into three broad categories, each having distinct design features and construction methods: (1) bored or mined tunnels; (2) cut-and-cover tunnels; and (3) immersed tube tunnels. These tunnels are commonly used for metro structures, high-way tunnels, and large water and sewage transportation ducts.

Bored or mined tunnels are unique because they are constructed without significantly affecting the soil or rock above the excavation. Tunnels excavated using tunnel-boring machines (TBM) are usually circular; other tunnels may be rectangular or horseshoe in shape. Situations where boring or mining may be preferable to cut-and-cover excavation include (1) significant excavation depths, and (2) the existence of overlying structures.

Cut-and-cover structures are those in which an open excavation is made, the structure is constructed, and fill is placed over the finished structure. This method is typically used for tunnels with rectangular cross-sections and only for relatively shallow tunnels (<15m of overburden). Examples of these structures include subway stations, portal structures and highway tunnels.

Immersed tube tunnels are sometimes employed to traverse a body of water. This method involves constructing sections of the structure in a dry dock, then moving these sections, sinking them into position and ballasting or anchoring the tubes in place.

This report is a synthesis of the current state of knowledge in the area of seismic design and analysis for underground structures. The report updates the work prepared by St. John and Zahrah (1987), which appeared in *Tunneling Underground Space Technol*. The report focuses on methods of analysis of underground structures subjected to seismic motion due to earthquake activity, and provides examples of performance and damage to underground structures during recent major earthquakes. The report describes the overall philosophy used in the design of underground structures, and introduces basic concepts of seismic hazard analysis and methods used in developing design earthquake motion parameters.

The report describes how ground deformations are estimated and how they are transmitted to an underground structure, presenting methods used in the computation of strains, forces and moment in the structure. The report provides examples of the application of these methods for underground structures in Los Angeles, Boston, and the San Francisco Bay Area.

This report does not cover issues related to static design, although static design provisions for underground structures often provide sufficient seismic resistance under low levels of ground shaking. The report does not discuss structural design details and reinforcement requirements in concrete or steel linings for underground structures. The report briefly describes issues related to seismic design associated with ground failure such as liquefaction, slope stability and fault crossings, but does not provide a thorough treatment of these subjects.[5] The reader is encouraged to review other literature on these topics to ensure that relevant design issues are adequately addressed.

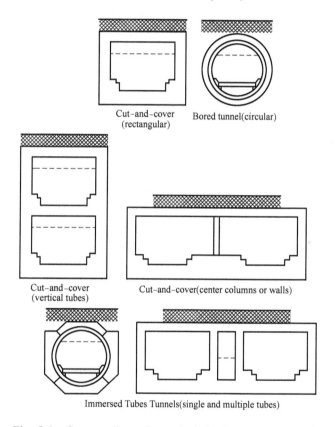

Fig. 8-1 **Cross sections of tunnels** (after Power et al., 1996).

8.3 Performance of Underground Facilities During Seismic Events

The following general observations can be made regarding the seismic performance of underground structures:

1) Underground structures suffer appreciably less damage than surface structures.

2) Reported damage decreases with increasing overburden depth. Deep tunnels seem to be safer and less vulnerable to earthquake shaking than shallow tunnels.

3) Underground facilities constructed in soils can be expected to suffer more damage compared to openings constructed in competent rock.

4) Lined and grouted tunnels are safer than unlined tunnels in rock. Shaking damage can be reduced by stabilizing the ground around the tunnel and by improving the contact between the lining and the surrounding ground through grouting.

5) Tunnels are more stable under a symmetric load, which improves ground-lining interaction. Improving the tunnel lining by placing thicker and stiffer sections without stabilizing surrounding poor ground may result in excess seismic forces in the lining. Backfilling with non-cyclically mobile material and rock-stabilizing measures may improve the safety and stability of shallow tunnels.

6) Damage may be related to peak ground acceleration and velocity based on the magnitude and epicentral distance of the affected earthquake.

7) Duration of strong-motion shaking during earthquakes is of utmost importance because it may cause fatigue failure and therefore, large deformations.

8) High frequency motions may explain the local spalling of rock or concrete along planes of weakness. These frequencies, which rapidly attenuate with distance, may be expected mainly at small distances from the causative fault.

9) Ground motion may be amplified upon incidence with a tunnel if wavelengths are between one and four times the tunnel diameter.

10) Damage at and near tunnel portals may be significant due to slope instability.

The following is a brief discussion of recent case histories of seismic performance of underground structures.

8.3.1 Underground structures in the United States

(1) Bay Area Rapid Transit (BART) system, San Francisco, CA, USA

The BART system was one of the first underground facilities to be designed with considerations for seismic loading. On the San Francisco side, the system consists of underground stations and tunnels in fill and soft Bay Mud deposits, and it is connected to Oakland via the transbay-immersed tube tunnel.

During the 1989 Loma Prieta Earthquake, the BART facilities sustained no damage and, in fact, operated on a 24 h basis after the earthquake. This is primarily because the system was designed under stringent seismic design considerations. Special seismic joints were designed to accommodate differential movements at ventilation buildings. The system had been designed to support earth and water loads while maintaining watertight connections and not exceeding allowable differential movements. No damage was observed at these flexible joints, though it is not exactly known how far the joints moved during the earthquake.

(2) Alameda Tubes, Oakland-Alameda, CA, USA

The Alameda Tubes are a pair of immersed-tube tunnels that connect Alameda Island to

Oakland in the San Francisco Bay Area. These were some of the earliest immersed tube tunnels built in 1927 and 1963 without seismic design considerations. During the Loma Prieta Earthquake, the ventilation buildings experienced some structural cracking. Limited water leakage into the tunnels was also observed, as was liquefaction of loose deposits above the tube at the Alameda portal. Peak horizontal ground accelerations measured in the area ranged between 0.1g and 0.25g. The tunnels, however, are prone to floatation due to potential liquefaction of the back-fill.

(3) L. A. Metro, Los Angeles, CA, USA

The Los Angeles Metro is being constructed in several phases, some of which were operational during the 1994 Northridge Earthquake. The concrete lining of the bored tunnels remained intact after the earthquake. While there was damage to water pipelines, highway bridges and buildings, the earthquake caused no damage to the Metro system. Peak horizontal ground accelerations measured near the tunnels ranged between 0.1g and 0.25g, with vertical ground accelerations typically two-thirds as large.

8.3.2 Underground structures in Kobe, Japan

The 1995 Hyogoken-Nambu Earthquake caused a major collapse of the Daikai subway station in Kobe, Japan. The station design in 1962 did not include specific seismic provisions. It represents the first modern underground structure to fail during a seismic event. Fig. 8-2 shows the collapse experienced by the center columns of the station, which was accompanied by the collapse of the ceiling slab and the settlement of the soil cover by more than 2.5m.

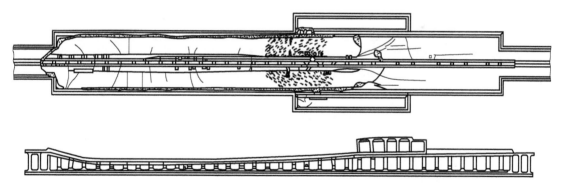

Fig. 8-2 Section sketch of damage to Daikai subway station.

During the earthquake, transverse walls at the ends of the station and at areas where the station changed width acted as shear walls in resisting collapse of the structure. These walls suffered significant cracking, but the interior columns in these regions did not suffer as much damage under the horizontal shaking. In regions with no transverse walls, collapse of the center columns caused the ceiling slab to kink and crack 150~250 mm wide appeared in the longitudinal direction.[6] There was also significant separation at some construction joints, and corresponding water leakage through cracks. Few crack, if any, were observed in the base slab.

Center columns that were designed with very light transverse (shear) reinforcement relative to the main (bending) reinforcement suffered damage ranging from cracking to complete collapse.

Center columns with zigzag reinforcement in addition to the hoop steel, as in Fig. 8-3, did not buckle as much as those without this reinforcement.

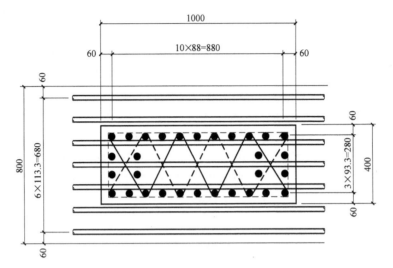

Fig. 8-3 Reinforcing steel arrangement in center columns.

It is likely that the relative displacement between the base and ceiling levels due to subsoil movement created the destructive horizontal force. This type of movement may have minor effect in a small structure, but in a large one such as a subway station it can be significant. The non-linear behavior of the subsoil profile may also be significant. It is further hypothesized that the thickness of the overburden soil affected the extent of damage between sections of the station by adding inertial force to the structure.[7] Others attribute the failure to high levels of vertical acceleration.

EQE (1995) made further observations about Daikai Station: Excessive deflection of the roof slab would normally be resisted by: ① diaphragm action of the slab, supported by the end walls of the station; and ② passive earth pressure of the surrounding soils, mobilized as the tube racks. Diaphragm action was less than anticipated, however, due to the length of the station. The method of construction (cut-and-cover, involving a sheet pile wall supported excavation with narrow clearance between the sheet pile wall and the tube wall) made compaction of backfill difficult to impossible, resulting in the tube's inability to mobilize passive earth pressures. In effect, the tube behaved almost as a freestanding structure with little or no extra support from passive earth pressure. However, it is not certain that good compaction would have prevented the structural failure of the column. Shear failure of supporting columns caused similar damage to the Shinkansen Tunnel through Rokko Mountain.

Several key elements may have helped in limiting the damage to the station structure and possibly prevented complete collapse. Transverse walls at the ends of the station and at areas where the station changed width provided resistance to dynamic forces in the horizontal direction. Center columns with relatively heavy transverse (shear) reinforcement suffered less damage and helped to maintain the integrity of the structure.[8] The fact that the structure was underground instead of being a surface structure may have reduced the amount of related damage.

A number of large diameter (2.0~2.4m) concrete sewer pipes suffered longitudinal cracking during the Kobe Earthquake, indicating racking and/or compressive failures in the cross-sections. These cracks were observed in pipelines constructed by both the jacking method and open-excavation (cut-and-cover) methods. Once cracked, the pipes behaved as four-hinged arches and allowed significant water leakage.

8.3.3 Underground structures in Taiwan, China

Several highway tunnels were located within the zone heavily affected by the September 21, 1999 Chi Chi earthquake ($M_L = 7.3$) in central Taiwan. These are large horseshoe shaped tunnels in rock. All the tunnels inspected by the first author were intact without any visible signs of damage. The main damage occurred at tunnel portals because of slope instability as illustrated in Fig. 8-4. Minor cracking and spalling were observed in some tunnel lining. One tunnel passing through the Chelungpu fault was shut down because of a 4m fault movement. No damage was reported in the Taipei subway, which is located over 100 km from the ruptured fault zone.

Fig. 8-4 Slope Failure at Tunnel Portal, Chi-Chi Earthquake, Central Taiwan.

8.3.4 Bolu Tunnel, Turkey

The twin tunnels are part of a 1.5 billion dollar project that aims at improving transportation in the mountainous terrain to the west of Bolu between Istanbul and Ankara. Each tunnel was constructed using the New Austrian Tunneling Method (NATM) where continuous monitoring of primary liner convergence is performed and support elements are added until a stable system is established. The tunnel has an excavated arch section 15m tall by 16m wide. Construction has been unusually challenging because the alignment crosses several minor faults parallel to the North Anatolian Fault. The August 17, 1999 Kocaeli earthquake was reported to have had minimal impact on the Bolu tunnel. The closure rate of one monitoring station was reported to have temporarily increased for a period of approximately 1 week, then became stable again. Additionally, several hairline cracks, which had previously been observed in the final lining, were being continuously monitored for additional movement and showed no movement due to the earthquake.[9] The November 12, 1999 earthquake caused the collapse of both tunnels 300m from their eastern portal. At the time of the earthquake, a 800m section had been excavated, and a 300m section of unreinforced concrete lining had been completed. The collapse took place in clay gauge material in the unfinished section of the tunnel. The section was covered with shotcrete (sprayed concrete) and had bolt anchors. Fig. 8-5 shows a section of the collapsed tunnel after it has been re-excavated. Several mechanisms have been proposed for explaining the collapse of the tunnel. These mechanisms include strong ground motion, displacement across the gauge material, and landslide.

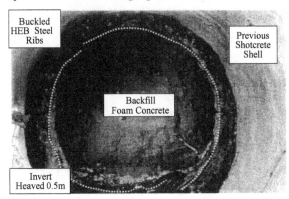

Fig. 8-5　Bolu Tunnel, re-mining of Bench Pilot Tunnels, showing typical floor heave and buckled steel rib and shotcrete shell.

8.3.5　Summary of seismic performance of underground structures

The Daikai subway station collapse was the first collapse of an urban underground structure due to earthquake forces, rather than ground instability. Underground structures in the US have experienced limited damage during the Loma Prieta and Northridge earthquakes, but the shaking levels have been much lower than the maximum anticipated events. Greater levels of damage can be expected during these maximum events. Station collapse and anticipated strong motions in major US urban areas raise great concerns regarding the performance of underground structures. It is therefore necessary to explicitly account for seismic loading in the design of underground structures.[10] The data show that in general, damage to tunnels is greatly reduced with increased overburden, and damage is greater in soils than in competent rock. Damage to pipelines (buckling, flotation) was greater than to rail or highway tunnels in both Kobe and Northridge. The major reason for this difference seems to have been the greater thickness of the lining of transportation tunnels. Experience has further shown that cut-and-cover tunnels are more vulnerable to earthquake damage than circular bored tunnels.

New Words and Expressions

infrastructure　*n.*　基础设施；基础建设

mound　*n.*　土堆，土丘；堆；土堤（mounded, 半埋设的）

liquefaction　*n.*　液化

geotechnical　*adj.*　岩土工程技术的

soil-plug　*n.*　土塞

solidify　*v.*　固化，（使）成为固体，（使）变硬，（使）变得坚固

hydropower　*n.*　水力发出的电力；水电

Unit 8　Introduction to Underground Engineering

chamber　洞室
overburden　*n.*　上部沉积（土），覆盖层，表土（层）
portal　*n.*　入口；桥门
intact　*adj.*　完整无缺的，未经触动的，未受损伤的
buckle　*v.*　（使）变形，弯曲
diaphragm　隔板
integrity　*n.*　整体，完整
backfill　*v.*　回填，充填
attenuate　*v.*　（使）变细；（使）变薄；（使）变小；减弱；减小
wavelength　*n.*　波长
sewage　*n.*　（下水道里的）污染；下水道；污水
excavation　*n.*　挖掘；开凿
ballasting　*n.*　压舱材料，道砟材料 *v.* 给……装上压舱物，给……铺道砟，使稳定（ballast 的现在分词）

Notes

1. Planners and the construction industry are preparing to apply competitive underground solutions to all types of infrastructure problems, including application of subsurface tunneling by means of shield-supported tunneling installations.

本句难点解析：这句话的主语是 planners and the construction industry，谓语是 are，宾语是 infrastructure problems。Including 引导的部分修饰 underground solutions，具体说明了地下解决方案。competitive，有竞争力的，这里可译为先进的。

本句大意如下：规划者和建筑行业正准备在所有类型的基础设施问题上采用先进的地下结构解决方案，包括采用盾构进行隧道安装的地下隧道应用方案。

2. Optimization of the cost level will require lump-sum turnkey contracts, with a clear agreement before contract award regarding uncertainties (which will be numerous when applying underground technology, especially during realization of the first such projects), risk control, quality control and performance monitoring.

本句难点解析：这句话的主语是 optimization，谓语是 will require，宾语是 turnkey contracts，with 引导定语修饰主句，which 引导从句修饰 uncertainties。lump-sum turnkey contracts，指交钥匙工程，即总包工程，一般是总价（闭口价）。

本句大意如下：成本层面的优化需要总包合同的总价，在确定合同前需要在不确定因素（应用地下技术时，特别是在第一个项目实施过程中有巨大的不确定因素）、风险控制、质量控制和性能监测方面达成明确共识。

3. The challenge to the engineering community and contractors in The Netherlands is to use these opportunities in such a way that Dutch industry can take a leading position in soft soil tunneling within a time frame of 10 years.

本句难点解析：这句话的主语是 challenge，谓语是 is，宾语是 opportunities，主语之后 to

引导的部分是用来修饰主语的。

本句大意如下：荷兰工程界和承包商面临的挑战是利用这些机会，使荷兰在未来十年内在软土隧道工程领域占据主导地位。

4. This report does not discuss pipelines or sewer lines, nor does it specifically discuss issues related to deep chambers such as hydropower plants, nuclear waste repositories, mine chambers, and protective structures, though many of the design methods and analyses described are applicable to the design of these deep chambers.

本句难点解析：这句话的主语是report，谓语是does not，宾语是pipelines or sewer lines，nor引导倒装句对主句进行补充说明，though引导条件状语从句。

本句大意如下：尽管很多设计方法和分析都被应用于这些深井洞室的设计，但本报告既不讨论管道或污水管道，也没有专门讨论与深井洞室有关的问题，比如水力发电厂、核废料库、矿井室和防护结构。

5. The report briefly describes issues related to seismic design associated with ground failure such as liquefaction, slope stability and fault crossings, but does not provide a thorough treatment of these subjects.

本句难点解析：这句话连接两个分句，主语是report，谓语分别是describes和does not，宾语是issues和treatment，related to引导定语修饰issues，associated with引导的定语从句修饰seismic design。

本句大意如下：该报告简要描述了与地基基础破坏有关的抗震设计问题，如液化、边坡稳定和断层交错，但没有提供彻底解决这些问题的方法。

6. During the earthquake, transverse walls at the ends of the station and at areas where the station changed width acted as shear walls in resisting collapse of the structure.

本句难点解析：这句话的主语是transverse walls，谓语是acted，宾语是shear walls，at引导的介词短语修饰主语，where引导的定语从句修饰areas。

本句大意如下：在地震过程中，站台两端和站台宽度变化区域内的横墙充当抵抗结构倒塌的剪力墙。

7. It is further hypothesized that the thickness of the overburden soil affected the extent of damage between sections of the station by adding inertial force to the structure.

本句难点解析：主语从句，这句话的形式主语是it，实际主语是that引导的部分，谓语是is，宾语是hypothesized，that引导部分中的主语是thickness。

本句大意如下：进一步假设，上覆土层的厚度能通过对结构施加惯性力，来影响站台各部分间的破坏程度。

8. Center columns with relatively heavy transverse (shear) reinforcement suffered less damage and helped to maintain the integrity of the structure.

本句难点解析：这句话的主语是center columns，谓语是suffered和helped，宾语是damage和integrity，with引导定语修饰主语。

本句大意如下：有相对较重的横向（剪力）钢筋的中心柱损伤较小，并且有助于维持结构的整体性。

9. Additionally, several hairline cracks, which had previously been observed in the final

Unit 8　Introduction to Underground Engineering

lining, were being continuously monitored for additional movement and showed no movement due to the earthquake.

本句难点解析：这句话的主语是 cracks，宾语是 additional movement 和 no movement，which 引导定语从句修饰主语。

本句大意如下：此外，先前在最终端的内衬中观测到的几处细微裂缝，也被持续地进行监测，以观察其进一步的移动，结果显示其未发生由于地震引起的移动。

10. Station collapse and anticipated strong motions in major US urban areas raise great concerns regarding the performance of underground structures.

本句难点解析：这句话的主语是 Station collapse and anticipated strong motions，谓语是 raise，宾语是 great concerns，regarding 引导的定语修饰宾语 concerns。

本句大意如下：在美国主要的城区，车站倒塌和预期的强烈振动引起了人们对地下建筑结构性能的广泛关注。

 **Exercises**

Translate the following phrases into Chinese.
1. shallow depth
2. negative skin friction
3. with respect to
4. immersed tube tunnel
5. tunnel-boring machine
6. a synthesis of
7. rock-stabilizing measure
8. Bay Area Rapid Transit (BART) system
9. tunnel lining
10. circular bored tunnel

Translate the following Sentences into Chinese.

1. In addition to the above-listed traditional types of positive drivers, there are new ones, which influence the choice in favor of underground alternatives.

2. Immersed tube tunnels are sometimes employed to traverse a body of water. This method involves constructing sections of the structure in a dry dock, then moving these sections, sinking them into position and ballasting or anchoring the tubes in place.

3. One tunnel passing through the Chelungpu fault was shut down because of a 4m fault movement. No damage was reported in the Taipei subway, which is located over 100 km from the ruptured fault zone.

4. The collapse took place in clay gauge material in the unfinished section of the tunnel. The section was covered with shotcrete (sprayed concrete) and had bolt anchors.

5. Each tunnel was constructed using the New Austrian Tunneling Method (NATM) where continuous monitoring of primary liner convergence is performed and support elements are added untile a stable system is established.

Unit 9

Introduction to Traffic Engineering

■ 9.1 Introduction

Traffic can be defined as the movement of pedestrians and goods along a route, and in the 21st century the biggest problem and challenge for the traffic engineer is often the imbalance between the amount of traffic and the capacity of the route, leading to congestion. Traffic congestion is not a new phenomenon. Roman history records that the streets of Rome were so clogged with traffic that at least one emperor was forced to issue a proclamation threatening the death penalty to those whose chariots and carts blocked the way. More recently pictures of our modern cities taken at the turn of the century show streets clogged with traffic. The dictionary describes 'traffic' as the transportation of goods, coming and going of persons or goods by road, rail, air, etc. Often in common usage we forget this wider definition and colloquially equate the word with motorised road traffic, to the exclusion of pedestrians and even cyclists. Traffic engineering is concerned with the wider definition of traffic.

In the introduction to his book Gordon Wells quoted the institution of Civil Engineering for his definition of traffic engineering, that is:

That part of engineering which deals with traffic planning and design of roads, of frontage development and of parking facilities and with the control of traffic to provide safe, convenient and economic movement of vehicles and pedestrians.[1]

The definition remains valid today but there has clearly been a change in the emphasis in the role of the engineer in the time. In the 1970s the car was seen as the future and the focus was very much 'predict and provide'. Traffic engineers were tasked with increasing the capacity of the highway system to accommodate what seemed and endless growth in motor traffic, often at the expense of other road users. Road capacity improvements were often achieved at the expense of pedestrian freedom of movement, pushing pedestrians to bridges and underpasses so that the surface could be given over to the car.[2] However, it is now generally, but by no means universally recognized that we will never be able to accommodate unconstrained travel demand by car and so increasingly traffic engineering has become focused on sharing space and ensuring that more sustainable forms of transport such as walking and cycling are adequately catered for.

This change has been in response to changes in both society's expectations and concerns about traffic and the impact of traffic on the wider environment. There has also been a pragmatic change forced on traffic engineers as traffic growth has continued unabated and so the engineer has been forced to fit more traffic onto a finite highway system.

Since 1970, road travel in the UK has increased by about 75% and, although many new roads have been built, these have tended to be inter-urban or bypass roads, rather than new roads in urban areas. Thus, particularly in urban areas, the traffic engineer's role is, increasingly, to improve the efficiency of an existing system rather than to build new higher capacity roads. Traffic engineering is used to either improve an existing situation or, in the case of a new facility, to ensure that the facility is correctly and safely designed and adequate for the demands that will be placed on it.

1) Highway Pavement Materials. Tar has been used for many years in road construction both as a binder for macadam and as a surface dressing initially on water-bound roads. Road tar is today specified in BS 761 by the following definition: 'Tar for use in road work is prepared entirely from crude tars produced wholly or substantially as a by-product in the carbonisation of coal at above 600 ℃ in externally heated retorts or coke ovens'.

The operation of proportioning the various aggregate sizes, adding binder and mixing the whole to produce as far as possible a homogeneous mass is a relatively simple process if a soft binder is employed; quality control is not important and only a small amount of mixed material is required. All that is required is a heated tray and several men with shovels.

2) Design of Concrete Pavements. Concrete pavements are constructed in a variety of forms by several different construction methods. They may be reinforced or unreinforced; if reinforced the steel may take the form of individual bars or welded mesh. The slabs may contain several different types of joint or they may be unjointed or continuous. Construction may be carried out by the conventional side-form process using a concreting train with many differing units, or one of several forms of slip-form paver may be employed operating with a minimum of additional equipment.

3) Flexible Pavement Thickness Design. The thickness design of highway pavements requires the following large number of complex factors to be considered.

① The magnitude and number of repetitions of the applied wheel loads and the contact area between the tyre carrying the load and the road surface.

② The stiffness, stability, durability, the elastic and plastic deformation and resistance to fatigue loading of the pavement layers.

③ Volumetric changes in the subgrade due to climatic changes, the deformation of the subsoil under load and the ability of the pavement layers to reduce the stress imposed on the subsoil by the wheel loads.

④ The severity and incidence of frost and rainfall.

4) Drainage. The importance of adequate drainage was realised by Roman road builders, but after the decline of the Roman Empire the standard of highway construction in Europe declined and planned drainage became almost non-existent. With the advent of industrialisation, increasing attention was paid to the removal of surface water and the lowering of the water-table beneath the pave-

ment. In the United Kingdom the pioneer road-building of Telford and Macadam laid stress on the incorporation of a camber or crown into the road so that surface water could be quickly removed. Increasing urban development made open ditches inappropriate and led to the increasing use of road gulleys connected to road sewers as the means of removing surface water. Improvements in the quality of road materials have resulted in impervious pavements, making the problem of sub-soil usually greatest in the cuttings, where cut-off drainage is required.

5) Earthworks. A site investigation is an essential first step in the design of any highway on a new location. The information obtained by the survey assists in locating the highway to avoid adverse geological conditions and in designing earthworks, pavement thickness, drainage works and bridge foundations.

6) Pavement Construction. With the drainage works completed and all services and ducts crossing the carriage-way in place, the formation may be sealed to protect the subgrade from excessive moisture changes. This is because a cohesive subgrade which has become excessively dry due to evaporation during a dry construction season may swell with subsequent increases in moisture content, resulting in differential movement of the pavement. On the other hand a subgrade that has become excessively wet is difficult to compact and to overlay with sub-base material. Universal sealing of the subgrade thus will not always be the answer because it may delay evaporation and hinder the evaporation from the soil. For work in the United Kingdom the following broad recommendations have been made. If the subgrade is formed during the months of October to April (inclusive), then it should be protected unless the sub-base can be laid and compacted on the same day as the subgrade is prepared. If the subgrade is cohesive and is prepared during May to September (inclusive), protection should be given if the formation is likely to be exposed for more than 4 days.

7) Pavement Maintenance. The twentieth century has seen a considerable improvement in the materials and constructional techniques used for highway pavements. This has resulted in a dramatic increase in the life of a pavement from the period when an annual surface dressing was necessary to maintain the shape of the pavement to the present time when design lives of from five to twenty years are common for heavily trafficked highways.

8) Parking Surveys. Every trip by a vehicle results in a parking act at the end of the trip. The importance of parking can perhaps be illustrated by the fact that, on average, a car in the UK is parked for about 23 hours a day. The vehicle may be parked on the street or off-street in a car/lorry/cycle park, or in a private garage. How vehicles arrive and depart from these parking places, how long they stay and under what circumstances define vehicular traffic and indeed some pedestrian traffic on the roads and help to determine what measures are required to meet or manage the demand. Therefore, it is very important to obtain an objective and unbiased understanding of this activity by properly constructed and conducted surveys.

9) Estimating Travel Demand. The estimation of travel demand is a fundamental part of traffic engineering design work. The key questions are how much effort needs to be expended in estimating demand and what method should be adopted. The answers depend on the nature of the design issues. For example, a minor traffic management design to improve road safety over a length of road in inner

London where traffic flows have been stable for many years will require little more than a survey of existing traffic. The reverse is true of a proposal for a new roadway to assist regeneration in an old urban area where design will depend on estimating the new traffic likely to be attracted to use the new road.

10) Capacity Analyses. The term capacity when referring to a highway link or junction is its ability to carry, accommodate or handle traffic flow.[3] Traditionally, capacity has been expressed in numbers of vehicles or passenger car units (PCU) (Vehicles vary in their performance and the amount of road space they occupy. The basic unit is the passenger car and other vehicles are counted as their PCU equivalent, e. g. a bus might be 3 PCUs and a pedal cycle 0. 1 PCU). In recent years public transport operators have applied pressure to consider highways in terms of their passenger-handling capacity and thus give a greater emphasis to the benefits of using high occupancy vehicles, such as buses or trams.

There is no absolute capacity value that can be applied to a given highway link, traffic lane or junction. Maximum traffic-handling capacity of a highway depends upon many factors including:

① The highway layout including its width, vertical and horizontal alignment, the frontage land uses, frequency of junctions and accesses and pedestrian crossings.

② Quality of the road surface, clarity of road marking, signing and maintenance.

③ Proportions of each vehicle type in the traffic flow and their general levels of design, performance and maintenance.

④ The numbers and speed of vehicles and the numbers of other road users, such as cyclists and pedestrians.

⑤ Ambient conditions including time of day, weather and visibility.

⑥ Road user levels of training and competence.

The capacity of a road junction is dependent upon many of the features that govern link capacity with the addition of the junction type, control method and vehicle turning proportions.

The expression 'level of service' when applied to a highway refers to the Highway Capacity Manual approach which defines a range of levels from the lowest which occurs during heavy congestion to the highest where vehicles can travel safely at their maximum legal speed.

11) Traffic Signs. An important part of any road is the means by which the traffic engineering conveys information about the road and any regulations that affects the way it is used to users. If this is done successfully it helps to make travel both safer and more efficient and it helps road users to ensure that they comply with the regulations governing the road that they are using.

Traffic signals are used to regulate and control conflicts between opposing vehicular or pedestrian traffic movements. Without the use of signals at some sites the major flow would dominate the junction, making entries from the minor road impossible or very dangerous. At other sites the minor road might interfere with the flow of major road traffic to such an extent that excessive congestion would occur. Traffic signals cannot only improve junction capacity, but can also improve road safety. The traffic engineer will need to know how best to provide parking, and how to control parking facilities both on-street and off-street, both in surface sites and structures.

9.2 Traffic Management and Control

Traffic management arose from the need to maximise the capacity of existing highway networks within finite budgets and, therefore, with a minimum of new construction. Methods, which were often seen as a 'quick fix', required innovative solutions and new technical developments. Many of the techniques devised affected traditional highway engineering and launched imaginative and cost-effective junction designs. Introduction of signal-controlled pedestrian crossings not only improved the safety of pedestrians on busy roads but improved the traffic capacity of roads by not allowing pedestrians to dominate the crossing point.

More recently the emphasis has moved away from simple capacity improvements to accident reduction, demand restraint, public transport priority, environmental improvement and restoring the ability to move around safely and freely on foot and by pedal cycle.[4]

There has been a significant shift in attitudes away from supporting unrestricted growth in highway capacity. The potential destruction of towns and cities and the environmental damage to rural areas is not acceptable to a large proportion of the population. Traffic management has, largely, maximised the capacity of the highway network yet demand and congestion continues to increase.

As traffic demand and congestion increased, drivers found alternative routes, often through residential areas. Road safety was compromised as drivers travelled at high speed to maximise the benefits of diverting from their normal route. Pressure from residents, in these areas, led to the introduction of area-wide Environmental Traffic Management Schemes (ETMS) during the 1970s and 1980s.

ETMSs attempted to deny these 'rat-runs' to queue-jumping traffic and to specific classes of vehicle such as wide or heavy vehicles. Many ETMSs were spectacularly effective and used such techniques as point road closures, physical width restrictions, one-way 'plugs', one-way streets and banned turns. These measures, which were designed to be restrictive for unwelcome traffic, often caused great inconvenience to residents, emergency vehicles and service vehicles. Frequently residents were prepared to tolerate severe inconvenience in order that a safe and tranquil environment could be restored.

1) Highway Layout and Intersection Design. It is unlikely that the traffic engineer alone will be required to design major highways. Highway design is a separate, albeit a closely related, discipline. At most the traffic engineer will provide preliminary layouts of access roads and intersections for developments and, therefore, a good understanding of basic highway design techniques and standards is needed.

Selection of an appropriate design speed may be considered the starting point for any scheme.[5] The DMRB Technical Standard TD 9/93 outlines the methods for selecting the link design speed.

In practice, the design speed for a particular route might be contained within a policy decision by the highway authority (HA). The engineer should consult the HA closely and, if no advice is forthcoming, suggest a method for determining a suitable design speed. In which case, the design speed should be selected by observation of the actual behaviour of the vehicles on the road in question.

Vehicle speeds are affected by many factors including speed limit, horizontal and vertical alignment, visibility, highway cross section, adjacent land use, spacing of junctions, accesses, pedestrian crossings and maintenance standards. The general condition and design of vehicles and driver ability, which changes over time can have a significant effect on vehicle speeds. It is usual to use a 85th percentile speed as the design speed (the speed below which 85% of drivers travel). The 85th percentile speed is determined from speed surveys using a radar speed meter or automatic traffic counters equipped to measure speeds.

In urban areas, where the speed limit is less than the national speed limit, the design speed is more likely to be based upon a policy decision. Frequently the traffic engineer is asked to reduce the speeds of vehicles to an acceptable level. In these circumstances it would be wholly inappropriate to use the 85th percentile speed as the design speed as this is often well above the existing or proposed speed limit. The engineer may be instructed to introduce measures that reduce or control 85%, or an even higher percentage, of vehicle speeds to the speed limit.

2) Road Safety Engineering. There are three factors that result in accidents:

① Road and environment deficiencies.

② Road user errors (human factors).

③ Vehicle defects.

Road and environment deficiencies account on their own for only 2% of all accidents but in combination with road user errors account for slightly less than 20%. Human factors on their own account for 75% of accidents.

Typical road and environment deficiencies are those which provide misleading visual information, or insufficient or unclear information to the road user. Only occasionally accidents are caused solely by bad design.

Human factors include excessive speed for the conditions, failing to give way, improperly overtaking or following too close and general misjudgement by both driver and pedestrian.

The two basic types of road accident, which by definition have to involve a vehicle, are:

① Personal injury.

② Damage only.

A personal injury accident (PIA) is an accident involving an injury. The PIA refers to the accident as the event, and may involve several vehicles and several casualties (persons injured). The accident must occur in the public highway (including footways) and become known to the police within 30 days of its occurrence. The vehicle need not be moving and it need not be in collision with anything. A casualty is a person killed or injured in an accident. <u>Casualties are subdivided into killed, seriously injured and slightly injured.</u>[6]

3) Traffic Calming. Traffic calming has two main objectives: the reduction in numbers of personal injury accidents and improvement in the local environment for people living, working or visiting the area.

Traditional traffic management uses physical measures and legislation to coerce, and mould driver behaviour to coax higher capacities out of the highway network, with improved levels of safety. Traffic calming now uses an expanded repertoire of measures and techniques to change driver's perception of an area.[7] Many streets portray the impression that they are vehicular traffic routes that have some other uses of lesser importance, such as shopping streets or for residential access. Traffic calming can alter the balance and impress upon the driver that the street is primarily for shopping or residential use and that vehicular traffic is of secondary importance.

Regardless of the prime cause of accidents, it has long been recognised that there is a direct correlation between accident severity and vehicle speed. Excessive speed for the prevailing road conditions can in itself be the prime cause of some accidents. Speeding traffic can cause severance effects between two parts of a community due to the difficulties experienced when pedestrians attempt to cross the road. High speed vehicles produce high noise levels and consequently degrade the environment. The Transport Research Laboratory (TRL) has identified that a 3%~7% reduction in accidents can be expected for every 1 kilometre/hour reduction in vehicle speeds in urban areas.

Increasingly traffic calming is being seen as part of the urban design toolkit. This means that traffic calming designs must be consistent with improvements to the urban streetscape. The traffic engineer must modify his designs to be sensitive to streetscape imperatives which will vary by the type of area. For example, the measures appropriate in an industrial area may well not be visually acceptable in a residential area or shopping street.

Almost from the dawn of the motor age, transport planners and policy makers have assigned a hierarchy to the road network with inter-urban trunk roads, primary distributors, district distributors, local distributors and access roads. In urban areas, increasing vehicular traffic levels and congestion has eroded the differences between the road types. Longer distance traffic has diverted to 'rat-runs' through local areas and traffic flows have grown on secondary roads to levels formerly associated with primary routes.

4) Public Transport Priority. Public transport priority has to be seen in the context of an overall urban transport strategy with objectives which include not only improved bus (or tram) operation and restraint of car-borne commuting but also an enhanced environment for residents, workers and visitors. Measures proposed must serve all these objectives and yet also be demonstrably cost-effective and enforceable.

Typical design objectives for public transport priority measures include:

① To improve the conditions and reliability of bus operations through the introduction of appropriate bus priority measures.

② To alter the traffic balance in favour of buses at those locations where this can be properly justified.

③ To improve conditions for bus passengers at stops and interchanges.

④ To improve road safety generally and, in particular, for pedestrians, cyclists and people with disabilities.

⑤ To review, where appropriate, hours of operation of waiting and loading restrictions.

⑥ To establish and implement the co-ordinated and coherent application of waiting, parking and loading enforcement regimes on bus route corridors.

⑦ To improve conditions for all road users and frontages on bus route corridors.

Achieving these objectives often involves compromises between improving bus operation and the needs of local businesses and residents for reasonable access and of pedestrians and cyclists for safe and convenient movement.

Bus priority measures should be seen as part of the tool kit that will enable the realisation of the transport strategy.[8] The impact of these measures on bus operation can be powerful, yet that impact should not be exaggerated. On their own, bus priority measures are unlikely to cause the major shift in travel from car to bus that is often needed to improve the urban environment. Yet, combined with other measures, bus priority can contribute to a strategy of improving the urban environment and road safety and minimising the need for car travel. Typical other measures include:

① A restrictive city centre parking policy for commuters.

② Improved bus services including park and ride.

③ Improved bus information for passengers.

④ More road space provided for pedestrians and cyclists.

5) Development Process and Sustainable Development. Most developments require planning consent from the local planning authority, that is, district or unitary authority, and the context for obtaining consent is described in legislation such as the Town and Country Planning Act 1990, in government policy guidance and in local authority plans. There are a few types of development which are controlled differently, for example mineral extraction which is determined at a county level, military works and major infrastructure such as strategic roads and airports.

6) Intelligent Transport Systems. Intelligent Transport Systems (ITS), also known as Transport Telematics, are concerned with the application of electronic information and control in improving transport.[9] We can already see some new systems implemented and can expect the pace of implementation to quicken. With a crystal ball, we can foresee how a typical journey to work may look in 10 years time.

7) Enforcement. In the past enforcement was, more or less, the exclusive preserve of the police service using the criminal law; however, as traffic levels and the number of offences have increased and police resources and priorities have changed, it has been recognised that the police service has become unable, or unwilling to devote sufficient resources to the enforcement of minor traffic offences. Further there has been a recognition that perhaps the structure and cost of the criminal, law process was perhaps inappropriate for the types of offences being dealt with. In simple words, the system was costing too much and was not able to cope with the increasing levels of minor offences being committed.[10]

8) Statutory Requirements. Traffic engineering activity, particularly on the highway, is controlled by an enormous number of laws and regulations which, taken together, set out:

① The rights of landowners.

② The rights and responsibilities of road users.

③ The rights and responsibilities of the highways authority and, by inference, the traffic engineer acting on behalf.

④ The powers of the police to enforce the law, particularly as it relates to traffic and travellers.

New Words and Expressions

capacity   *n.*   容量，承载能力
congestion   *n.*   堵车；阻塞
traffic engineering   交通工程
tar   *n.*   沥青；柏油
macadam   *n.*   铺路用的碎石料
carbonisation   *n.*   碳化
concrete pavements   混凝土路面
earthworks   *n.*   土方（工程）
maintenance   *n.*   保养维护
highway layout   公路布局，公路规划
planning authority   规划局
telematics   *n.*   信息技术
innovative   *adj.*   革新的；创新的
sustainable development   可持续发展
pedestrian   *n.*   行人

Notes

1. That part of engineering which deals with traffic planning and design of roads, of frontage development and of parking facilities and with the control of traffic to provide safe, convenient and economic movement of vehicles and pedestrians.

本句难点解析：design 后并列三个 of 都在讲 Gordon Wells 这本书中设计的内容。That part 指前文的交通工程研究。

本句大意如下：交通工程研究的主要内容是交通规划和道路、停车设施的设计，及控制交通以保障车辆和行人安全、方便、经济的通行。

2. Road capacity improvements were often achieved at the expense of pedestrian freedom of movement, pushing pedestrians to bridges and underpasses so that the surface could be given over to the car.

本句难点解析：movement 后为从句，解释道路通行能力的改善，牺牲了行人的移动自由。

本句大意如下：道路通行能力的改善常以牺牲行人的通行自由为代价，迫使行人从天桥或地下通道通行以给汽车让出路面。

3. The term capacity when referring to a highway link or junction is its ability to carry, accommodate or handle traffic flow.

Unit 9　Introduction to Traffic Engineering

本句难点解析：when 从句修饰前面的容量。

本句大意如下：高速公路交接处的容量是指运载、容纳和处理交通流量的能力。

4. More recently the emphasis has moved away from simple capacity improvements to accident reduction, demand restraint, public transport priority, environmental improvement and restoring the ability to move around safely and freely on foot and by pedal cycle.

本句难点解析：to 后并列多个名词说明现在的研究重点。

本句大意如下：最近的研究重点已经从简单的改善通行能力转变为减少事故发生，如抑制需求、优先采用公共交通、改善环境以及恢复行人和自行车自由安全通行。

5. Selection of an appropriate design speed may be considered the starting point for any scheme.

本句难点解析：starting point 是起点，要点的意思。

本句大意如下：设计速度的合理选择是任何方案中首先考虑的要点。

6. Casualties are subdivided into killed, seriously injured and slightly injured.

本句难点解析：casualties 是"伤亡"的意思。这句话主语是 casualties，be subdivided into 是被分成的意思。

本句大意如下：伤亡可分为死亡、重度受伤和轻伤。

7. Traffic calming now uses an expanded repertoire of measures and techniques to change driver's perception of an area.

本句难点解析：这句话主语是 traffic calming，结构是 use sth. to do sth.。

本句大意如下：如今，交通稳静化使用更多的方法和技术去改变驾驶员对一个地区的感知。

8. Bus priority measures should be seen as part of the tool kit that will enable the realisation of the transport strategy.

本句难点解析：that 后的从句修饰前面的 tool kit，说明其作用。

本句大意如下：公交优先措施应当被视为能帮助实施交通策略的措施之一。

9. Intelligent Transport Systems (ITS), also known as Transport Telematics, are concerned with the application of electronic information and control in improving transport.

本句难点解析：这句话主语是 ITS 谓语为 are，also known as 从句修饰 ITS 解释其另一种说法。

本句大意如下：智能交通系统，又被称为交通远程信息处理，是一个利用电子信息和控制技术来改善交通的系统。

10. In simple words, the system was costing too much and was not able to cope with the increasing levels of minor offences being committed.

本句难点解析：这句话的主语是 system，两个谓语动词 was 并列。

本句大意如下：简单来说，这个系统成本太高且不能处理逐渐增多的轻微犯罪行为。

 Exercises

Translate the following phrases into Chinese.

1. highway pavement
2. in the role of
3. accommodate unconstrained travel demand
4. adequate for the demands
5. signal-controlled pedestrian crossings
6. highway network
7. automatic traffic counters
8. speed limit
9. caused solely by bad design
10. demonstrably cost-effective and enforceable

Translate the following sentences into Chinese.

1. Traffic can be defined as the movement of pedestrians and goods along a route, and in the 21st century the biggest problem and challenge for the traffic engineer is often the imbalance between the amount of traffic and the capacity of the route, leading to congestion.

2. The definition remains valid today but there has clearly been a change in the emphasis in the role of the engineer in the time. In the 1970s the car was seen as the future and the focus was very much 'predict and provide'.

3. Tar has been used for many years in road construction both as a binder for macadam and as a surface dressing initially on water-bound roads.

4. Road safety was compromised as drivers travelled at high speed to maximise the benefits of diverting from their normal route.

5. In practice, the design speed for a particular route might be contained within a policy decision by the highway authority (HA).

Unit 10

Hydraulic Engineering

10.1 Introduction

The major function of a hydraulic project (i. e., water project) is to alter the natural behavior of a water body (river, lake, sea, groundwater) by concentrating its flow fall.[1] It is intended for purposeful use for the benefits of national economy and to protect the environment, including electric power generation, flood control, water supply, silt mitigation, navigation, irrigation and draining, fish handling and farming, ecologic protection, and recreation. It is common that a number of hydraulic structures (i. e., hydraulic works) of general or special purposes are constructed to form a single or integrated hydraulic project to comprehensively serve foregoing purposes. Such a project is known as the water resources project or hydropower project in China, and the latter is primarily for electric power generation in addition to other possible benefits. The general-purpose and special-purpose hydraulic structures which are parts of a hydraulic project can be further divided into main, auxiliary, and temporary structures.

10.2 Types of Hydraulic Structures

Hydraulic structures are submerged or partially submerged in water. They can be used to divert, disrupt, or completely stop the natural flow.[2] Hydraulic structures designed for integrated river, lake, or seawater projects are referred to river, lake, and marine water works, respectively. By the features of their actions on the stream flow, main hydraulic structures are distinguished as water retaining structures, water conveying structures, and special-purpose structures.

10.2.1 Water Retaining Structures

Dams (inclusive barrages) are typical water retaining structures that affect closure of the stream and create heading-up afflux.[3] Made of various materials, dams fall into soil and/or rockfill embankment, concrete, reinforced concrete, masonry, and wooden. Of which, the first two are the most prevalent nowadays.

(1) Concrete Dams

By the structural features, concrete dams are termed as gravity (massive), buttress, and arch.

A gravity dam is a concrete structure resisting the imposed actions by its weight and section without relying on arch. In its common usage, the term is restricted to solid masonry or concrete dam which is straight or slightly curved in plan. The downstream face of modern gravity dam is usually of uniform slope which if extended would intersect the upstream face at or near the maximum reservoir level. The upstream face is normally of steep uniform slope or vertical with a steep batter (flared) near the heel. The upper portion is thick enough to resist the impact of floating debris and to accommodate a roadway and/or a spillway. The thickness of section at any elevation is adequate to resist sliding and to ensure compressive stresses at the heel under different loading conditions.

A buttress dam depends principally upon the water weight in addition to the concrete weight for stability. It is composed of two major structural elements: a water-supporting deck of uniform slope and a series of buttresses supporting the deck. Buttress dams are customarily further classified according to the deck type: A flat slab dam is one whose deck comprises flat slabs supported on the buttresses; a multi-arch dam consists of a series of arch segments supported by buttresses; and a massive-head buttress dam is formed by flaring the upstream edges of the buttresses to span the spaces between the buttresses.

An arch dam is always curvilinear in plan with its convex side facing headwater. On its vertical cross section, the dam is a relatively thin cantilever slightly curved. An arch dam transmits a major part of the imposed actions to the canyon walls mainly in the form of horizontal thrusts from it abutments.

By the water flow features, concrete dams may be non-overflow and overflow, and the latter releases water through openings (outlets) that can be free overflow and/or submerged under the headwater level, i.e., orifices, deep openings, and bottom outlets.

(2) Embankment Dams

Embankment dams are massive fills of natural ground materials composed of fragmented particles, graded and compacted, to resist seepage and sliding. The friction and interlocking of particles bind the material particles together into a stable mass rather than by the use of a cementitious substance (binder). All modern embankment dams have basically trapezoidal cross section with straight or broken contour of upstream and downstream slopes. The topmost edge of the slope is the crest, and the lowermost edge of the slope is the toe or heel. Horizontal portions on the dam slope surfaces are termed as "berms".

The actions of the impounded reservoir create a downward thrust upon the mass of the embankment, greatly increasing the pressure of the dam on its foundation, which in turn adds force effectively to seal and make the underlying dam foundation waterproof, particularly at the interface between the dam and its streambed.

There are various types of embankment dams. On the basis of the natural ground materials used, earthfill dams are compacted by fine-grained soils accounted for over 50% of the whole placed volume, whereas rockfill dams are compacted by coarse-grained materials accounted for over 50% of the whole placed volume. On the basis of section zoning, homogenous and zoned embankment dams may be distinguished, and among the latter category, the term "decked rockfill dams" is used particularly for the rockfill dams employing thin upstream membrane of nonnatural materials such as asphaltic concrete, reinforced concrete, and geomembrane.

Embankment dams are commonly constructed as non-overflow. However, small rockfill dams are occasionally allowed for being topped over the crest if the dam crest and downstream face are adequately reverted.

10.2.2 Water Conveying Structures

Water conveying structures are artificial channels cut in the ground and made of either ground materials such as soil and rock (e.g., canals and tunnels) or artificial materials such as concrete and metal (e.g., aqueducts, flumes, siphons, pipelines).[4]

10.2.3 Special-purpose Hydraulic Structures

Special-purpose hydraulic structures are accommodated in a hydraulic project to meet the requirements of:

1) Hydroelectric power generation, inclusive power plants, forebays and head ponds, surge towers and shafts, etc.

2) Inland waterway transportation, inclusive navigation locks and lifts, berths, landings, ship repair and building facilities, timber handling structures and log passes, etc.

3) Land reclamation, inclusive sluices (head works), silt tanks, irrigation canals, land draining systems, etc.

4) Water supply and waste disposal (sewerage), inclusive water intakes, catchment works, pumping stations, cooling ponds, water after treatment plants, sewage headers, etc.

5) Fish handling, inclusive fish ways, fish locks and lifts, fish nursery pools, etc.

10.3 Layout of Hydraulic Projects

A hydraulic project is usually huge and comprises several hydraulic structures erected over an extensive geographic area and is intended to serve large-scale social and economic program. The hydraulic project with retaining structures built on the main stream of a river or canal, which creates afflux head of water, is known as operating under head; otherwise, it would be a headless (non-head) water project. Fig. 10-1, Fig. 10-2, and Fig. 10-3 show three typical water resources and hydropower projects in China.

Fairly number of alternative layout schemes are compared with respect to technology and economy aspects before the final optimal layout scheme is decided, which should facilitate the construction and the management as well as reduce the investment, on the premise of ensuring project safety.[5] The key issues facing the layout of a hydraulic project are as follows:

(1) Flood release during service period and river diversion during construction period commonly dominate the project layout. The spillway should attain sufficient discharge capacity;

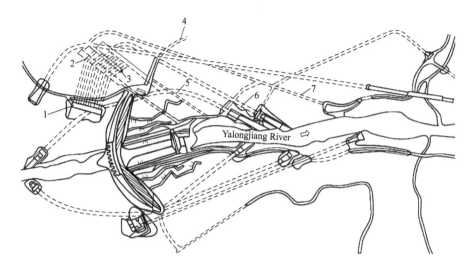

Fig. 10-1 Ertan Hydropower Project (arch dam)—China.
1—Power intake 2—Main machine house 3—Main transformer chamber 4—Left bank high way
5—Number 2 tail race tunnel 6—Number 1 tail race tunnel 7—Access tunnel for power house

meanwhile, its detrimental effects such as downstream scouring, silt depositing, and disturbance to power plant operation should be avoided or alleviated. The possibility of the reconstruction and reuse of temporary diversion structures as part of permanent flood or silt releasing structures is taken into account in the project layout, too.

(2) Power plant should be located at places of easy to access and outgoing power lines. Concrete dam projects usually install power plant at dam toe. Bank power plant or underground power plant is also frequently employed under the situation of narrow river rapids or for the purposes to avoid construction interference and/or to collect more head.

(3) Navigation structures (e.g., ship lock and lift) are located on the one or on the both riverbanks, where the flow conditions are advantageous for the upstream and downstream approach and berth of vessels. To avoid the interferences in the construction and operation, navigation structures are best to be separated with the intakes and the tailrace channel of the power plant.

(4) Silt flushing structures should be layout carefully with respect to the reservoir/pool sedimentation process and the silt protection requirement of the project, of which the intakes of power plant and canal head works are particularly emphasized. It is customarily to provide certain sized bottom outlets as desilting sluices (excluders) for flushing silt, to keep effective reservoir storage capacity and free of silt in front of intakes. The solid gravity dams and arch dams may meet this requirement by providing bottom outlets with low intake elevation and large orifice size. On the contrary, tunnels for releasing flood inflow and flushing silt are demanded for the projects using embankment dams.

(5) Dam safety is the most important. Particularly, the serious structural and equipment accidents as well as the civil air defense of dams and power plants should not be overlooked. The major countermeasure is drawing down or emptying reservoirs in time. To meet this requirement, in addi-

tion to crest spillway, intermediate and bottom outlets have to be provided to control the reservoir water level as flexibly as possible.

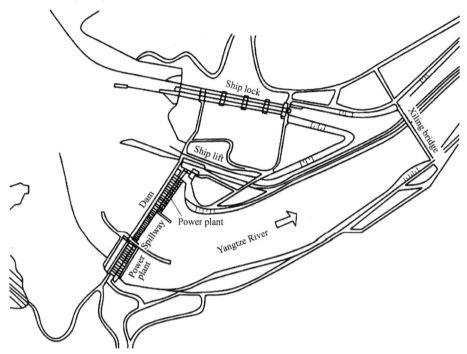

Fig. 10-2 Three Gorges Water Resources Project (gravity dam) —China.

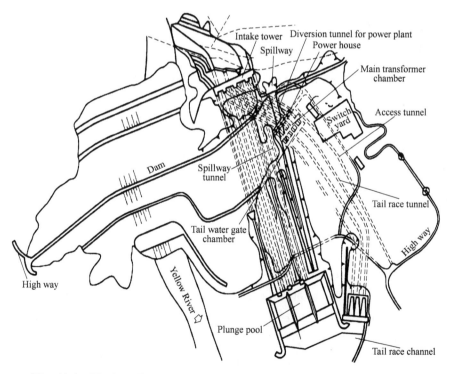

Fig. 10-3 Xiaolangdi Water Resources Project (embankment dam) —China.

Viewing from the high dams and large power stations in operation or under construction in China, the project layout is often, although not always, focused on flood releasing and powerhouse locating, which may be grouped into several representatives with respect to major dam types.

10.4 Classification of Hydraulic Projects and Their Design Safety Standards

In the first step, water resources and hydropower projects are distinguished as 5 classes in China according to their scales, benefits, and importance in the society and national economy. Secondly, permanent hydraulic structures in the project are classified into 5 grades according to their importance in the project concerned. The higher the structure grade, the higher the design safety standard is stipulated, regarding:

1) Flood-resisting ability, e.g., flood standard (recurrence interval), freeboard height, etc.

2) Earthquake-resisting ability, e.g., seismic design standard, earthquake countermeasures, etc.

3) Strength and stability, e.g., material strength, allowable safety factor against sliding, crack prevention requirements, etc.

4) Material type, quality, durability, etc.

5) Operation reliability, e.g., margin of structural size, monitoring instrumentation, etc.

There are two state standards in China for the classification of water resources and hydropower projects with corresponding design safety standards, due to the organization reshuffling history of the state central government: One is mainly adopted by the design institutes belong to the formal Ministry of Water Resources (with entry word of SL), and another one is mainly adopted by those belong to the formal State Economic and Trade Commission (with entry word of DL).

The design reference period of permanent retaining structures with grade 1 is stipulated as 100 years, while that of the other permanent structures is 50 years. For particular large-scale projects, the design reference period of permanent retaining structures must be decided after special studies; the design reference period of temporary structures may be decided according to their anticipated service time plus possible construction delay.

DL 5180—2003 *Classification & design safety standard of hydropower projects* also recommends the structural reliability theory for the design of hydraulic structures, and the target reliability index is stipulated to represent the design safety standard. However, where the structural reliability design is not available due to the delay in the design code updating for a certain structure, the determinant design method may have to be employed, and under such circumstances, the allowable safety factor is stipulated to represent the design safety standard. In the application of structural reliability theory, the structural grades 1~5 are corresponding to the structural safety grades I, II, and III, respectively.

10.5 Water Resources and Hydropower Resources in China

Be situated in the southeast part of the Eurasian continent on the west coast of the Pacific Ocean, adjacent to the Himalayas Mountain known as the "Roof of the World" in the southwest and to the Siberia and the Mongolian Plateau in the north, China has a complex and diversified landscape varying greatly in altitude. The Qinghai—Tibet Plateau in the west of the country is a vast area with the highest altitude in the world. In the north of the Qinghai—Tibet Plateau, there are mountain systems of Altai, Qilian, Daxing'an, etc., and in its south, there are mountain ranges of Himalayas, Hengduan, Wuyi, etc. A great number of rivers originate on these plateaus and mountain systems and flow across the country. These mountains and rivers determine the basic features of the state landscape and create superior natural conditions for the two major elements of hydropower resources: runoff and fall.[6] The theoretical and technically exploitable as well as the economically exploitable hydropower resources in China are all ranked at the first place in the world.

There are seven main river systems in China, namely the Yangtze River, the Yellow River, the Pearl River, the Huaihe River, the Haihe River, the Liaohe River, and the Songhuajiang River. In addition, there are also the southeast coastal river systems including the Qiantangjiang River and the Minjiang River, the northeast international river systems including the Heilongjiang River and the Yalujiang River, and the southwest international river systems including the Lancangjiang River and the Nujiang River and the Yaluzangbujiang River, the northwest international rivers of the Ertix and the Ili. There are also several interior river systems of Tarim and others in the provinces or autonomous regions of Xinjiang, Gansu, Inner Mongolia, and Qinghai.

According to the statistics data, there are more than 50000 rivers in China with a drainage area of over 100km^2. Among them, there are more than 1600 rivers with a drainage area of over 1000km^2, 20 rivers with a length of over 1000km, and 3019 rivers with theoretical hydropower resources of over 10MW.

The average annual precipitation in China is 648mm, i.e., about 6190 billion m^3 in total. Accordingly, the total amount of annual river runoff in China is about 2711 billion m^3. There are 17 rivers having an annual runoff over 50 billion m^3, which is ranked at the fifth place in the world after Brazil, Russia, Canada, and USA.

According to the data provided by the State Development and Reform Commission in 2004 (National Leading Group for the Re-check of the National Hydropower Resources Survey 2004), the theoretical hydropower resources in the Chinese mainland are 695GW, corresponding to an annual energy output of 6083TW · h; the technically exploitable hydropower resources are 542GW, which are at the first place in the world. Accordingly, China has planned 270 hydropower power stations with installed generator capacity over 300MW, of which 100 are over 3000MW and 800 are between 50MW and 300MW.

Due to the geographic features of the country, scarce cultivable land and huge population are highly concentrated in the eastern China, while 82.9% of water and hydropower resources are distributed in the vast western part of the country. The society development, particularly in the middle and lower river reaches with huge population, is often encountered with difficulties to satisfy multiple purposes such as flood control, agriculture irrigation, navigation, water supply, aquaculture, recreation, and ecology and is liable to conflicts among those purposes to some extent in terms of quantity and quality of water consumption, as well as spatial-time allocation. Therefore, China is actually a country in short of water resources: The average water resources per capita is about 2100m^3 and per hectare is about 22500m^3, which is only 30% and 50% of the world's level, respectively. In order to give priority to the full use of clean and renewable hydropower resources and meet the electric power demands, the high-quality management, rational exploitation, careful protection, and sustainable utilization of water resources are and will continue to be a strategically paramount issue in the economical and social development of the country.

■ 10.6 Hydraulic Engineering in China

Hydraulic engineering in China has gained rapid progress since the 1950s until the end of the 2000s, as has been interspersed foregoing. In the twenty-first century, the hydropower exploitation in China will attain a greater breakthrough by two steps.

The first step was from 2001 to 2010. By the end of 2010, the national hydropower capacity reached 210000MW, and the corresponding annual energy output was 650 TW · h. The hydropower accounted for about 30% in the national electric power supply. The 5200 dams higher than 30m were completed or under construction, of which 145 exceeded 100m. Among the 450 completed hydroelectric stations larger than 50MW in generator installation capacity (including 21 pumping storage stations), 100 exceeded 300MW (including 15 pumping storage stations) and 40 were of especially large scale whose installation capacity exceeded 1000MW (including 7 pumping storage stations). By the year of 2010, the total reservoir storage capacity reached 1/6 of the river annual runoff of the whole country, which played important role in the flood protection, irrigation, and water supply by covering 0.35 Billion population and 0.033 Billion ha farmland as well as hundreds of large to medium cities including Beijing, Tianjin, Guangzhou, Shanghai, and Wuhan. To meet the development requirements of metropolis cities, a large number of long distance water diversion projects were constructed, and more than 100 large to medium cities relied mainly on the reservoir supply for domestic living and industry, such as the Miyun reservoir for Beijing, the Panjiakou reservoir for Tianjin, and the Shenzhen reservoir for Shenzhen and Hongkong.

The second step is from 2011 to 2050. During this period, under the state policy guidance of exploiting her vast western area, China will nearly complete the exploitation of her hydro-energy potential, and there will be tens of super-large-scale hydropower plants built in the western China. Specially, the Motuo hydropower project, with design installed generator capacity larger than 40000MW, will be put into operation. It will become the greatest base of the hydropower in the world. The capacity of electricity transferred from the West to the East will exceed 150000MW.

Following the progress in the construction of large-scale water resources and hydropower projects, China has reached international level concerning the research, design, and construction technologies.[7] The largest hydraulic project completed in the world—the Three Gorges project—provides electric capacity of 22400MW; the world's highest arch dam (Xiaowan, $H=294.5$m), the world's highest concrete-faced rockfill dam (Shuibuya, $H=233$m), and the world's highest RCC gravity dam (Longtan, $H=216.5$m) are all erected in China.

10.7 Purposes of Hydraulic Projects

A hydraulic project may be small or large, simple or complex, and single or multiple purposed, and it should provide the functions to accomplish the optimum development of related water and hydropower resources.

In many cases, the project will be multi-purposed. For this reason, the investigations may comprise a large number of matters, and some or all of them will influence the selection of the project site and scale.[8] Hence, the entire project must be investigated as a whole before the design requirements for each single structure, such as the dam, can be firmly established.

Main aspects of the river development using hydraulic projects, with particular emphasis upon the design requirements for dams and reservoirs, will be presented hereinafter.

10.7.1 Flood Control

Many river basins are frequently suffered from destructive floods and are, therefore, difficult to be used for farming and residence. A very effective way for flood defense is to build hydraulic projects with capacious reservoirs. Many projects (e.g., the Three Gorges Project) owe a good deal or majority of their social and economic benefits to flood control function. Sometimes, hydraulic projects are specially built to fight floods only.

In the design of flood control projects and structures, it should be taken into account that:

1) The relation of the cost for flood control with the benefits to be derived through the reduction of cumulative damage should be favorable and in light of public interest, as compared to alternative means for obtaining similar benefits.

2) The temporary storage for design and check floods must be sufficient to cut the major peak inflows or to lower down the frequency of minor floods.

3) Flood control must be effective and reliable, and so far as it is predictable, the method of flood control should be automatic rather than manual.

10.7.2 Irrigation

Nowadays, about 230 million hectares of farmlands are irrigated all over the world—as much as 70% of the water taken from rivers is used for irrigation, of which 75% never returns back to the streams. In some regions, especially in the Middle East, the Central Asia and northern China, farming without irrigation would be unfeasible at all.

It is customary to distinguish between gravity irrigation and pumping irrigation. The former depends on the head created by dams over the elevation of the farmlands to which the water is delivered. The latter employs powered pumps, so the water can be lifted to any desired elevation, to be distributed later over the ramified system comprised of canals and conduits.

The desired amount of water is stored in reservoirs, and the power for pumps is furnished by hydropower plants that are generally components of multi-purposed hydraulic projects. Most modern China's irrigation systems are supported by multi-purposed hydraulic projects.

For successful irrigation, the supply of water must be adequate at an economically reasonable capital investment per unit of area and must be easy for operation and maintenance.

10.7.3 Power Generation

It is a sad fact that thermal power stations, especially those of coal-burned types, discharge a lot of ash and noxious gases into the atmosphere and foul up large territories. Of these, surplus dioxide is fraught with the gravest consequences. The thing is that it mixes well with water vapors and yields sulfuric acid, which, upon precipitation, poisons water bodies. In recent years, the toxic haze phenomenon frequently occurs in a wide territory across the northern and eastern China, which is mainly blamed for this coal-burned pollution, apart from another major pollution sources from steel and cement industries, city infrastructure construction, and vehicle exhaust.

Although they do not pollute the atmosphere, nuclear power plants present another danger since they produce radioactive wastes that are rather difficult to get rid of. The Fukushima Daiichi nuclear disaster caused by the Tohoku earthquake induced tsunami on March 11, 2011, gives a global warning for the potentiality due to uncontrolled consequences of nuclear power plant accident.

Hydropower stations are sufficiently "clean" enterprises that result in few soil, water, or air pollution. Therefore, they may considerably reduce the overall pollution compared to thermal or nuclear power stations. As far as the pollution problems are concerned, hydropower is extremely attractive.

On a global scale, the unlimited and uncontrolled development of power engineering (nuclear power plants being the most important cause for concern) may, in the long run, upset the thermal balance of the Earth—a fact its consequences to the mankind are very difficult to predict insofar. In this aspect, hydraulic power engineering, which ultimately depends on solar energy (in fact, it merely redistributes the energy released by the Sun) and therefore does not affect the thermal balance of the planet, appears to be an ideal choice.

Where the power generation is targeted in the development of a hydraulic project, the capacity of the power generating equipment and the load demand are closely related to the quantity of water available and the amount of storage provided, which in turn, dictate the height of the dam.

10.7.4 Navigation

River transportation plays an important role in water economy. As a rule, inland water transportation capacity will be raised considerably by building dams and barrages. The point is made that a chain of storage reservoirs improves navigation depths, straightens navigation channels, and ensures the pathway of large ships. Also, the regulated outflow from reservoirs improves navigation conditions in the downstream reaches. In China, most inland waterways depend on major hydraulic projects. Examples are the Yangtze River and the Yellow River waterways.

When a dam is constructed on a large river considering upstream and downstream navigation, it may be desirable to construct ship locks or lifts to provide pathway for vessels over the barrier. Depending upon topographic and geologic conditions, the locks may be the integral components of the dam or entirely separate structures. The functional design criteria are the dimensions of lock chamber or handling trough and the draft of vessels to be accommodated, and the estimated number of vessels passing upstream and downstream at peak periods without excessive delay.

10.7.5 Domestic and Municipal Purposes

Much water is consumed by metallurgical, chemical, wood pulp, and paper industries. Among major industry users are also thermal and nuclear power plants. In many industrially developed regions of China, the demand for water cannot be met by the local water resources solely. To deal with the problem, reservoirs are built to store the most of the local runoff, and large water developments are set up to divert water from other basins. Most of these water supply systems originate from the reservoirs of multi-purposed hydraulic projects.[9]

Although it constitutes merely about one-tenth of the industrial water consumption, public water supply, which meets the immediate needs of the population, should be taken into account seriously. Consequently, it is hardly surprising that the primary function of quite a number of multi-purposed hydraulic projects is to supply water for domestic daily consumption.

Sometimes, there is a need for the hydraulic project in a region where stream flow either ceases entirely or is reduced to extremely low levels during seasons of the year, and where such natural stream flow is the principal source of water supply for one or more communities, water storage creation for stream flow regulation may be justified apparently.

The quality of the water must be such that it can be rendered portable and usable for domestic and most industrial purposes by economical treatment methods. It should meet state public health standards with regard to bacterial purity, taste, color, odor, and hardness. Control and protection of watershed areas are desirable for municipal water supply reservoirs.

10.7.6 Environment Protection

It might be a selfish but reasonable idea that the important environmental issues are those that the most direct concern to the livelihood and well-being of mankind, and the various other living things are of concern to human to whatever degree their existence is important to his living conditions, i.e., there is a close relation between the important environmental issues and the social needs of human beings.

Among the various beneficial environmental—social effects of hydraulic projects, they may be distinguished as farmland improvement by irrigation, higher standard in flood protection, enhanced water quality and supplying ability for domestic and municipal uses, clean power supply without consumption of fuel, and fishery and recreational development. These benefits, partially measurable in economic terms, are among the principal objectives of a hydraulic project. <u>However, the negative impacts of the hydraulic project on the environment should be never overlooked in the design and construction.</u>[10]

10.7.7 Recreation and Other Purposes

A reservoir might significantly make an excellent site for various recreational facilities and health farms. Sometimes, small reservoirs are built specially for recreation purposes.

Occasionally, a hydraulic project is proposed to regulate the water level in shallow lakes and swamps other than those purposes heretofore enumerated. For example, the project for the detention or diversion of stream flow to conserve it by transforming surface water to groundwater through the process of infiltration could be planned and constructed. To justify the project economically, however, it must be determined that the soil characteristics will permit infiltration to occur in a desirable quantity.

New Words and Expressions

hydraulic *adj.* 水力的
electric power generation   发电
hydraulic structures   水工结构，水工构筑物
reinforced concrete   钢筋混凝土
gravity dam   重力坝
buttress dam   支墩坝
embankment dams   土石坝
seepage *n.* 渗流
alleviate *v.* 减轻，缓和；*n.* 减轻，缓解
silt flushing structures   淤泥冲刷结构
reliability *n.* 可靠度
runoff *n.* 径流
hydropower resources   水利资源
reservoirs *n.* 水库
cumulative *adj.* 累积的
dictate *v.* 控制
topographic *adj.* 地形（学）的
metallurgical   冶金（学）的

Unit 10　Hydraulic Engineering

1. The major function of a hydraulic project (i. e., water project) is to alter the natural behavior of a water body (river, lake, sea, groundwater) by concentrating its flow fall.

本句难点解析：这句话的主语是 function，谓语是 is。

本句大意如下：水利工程的主要功能是通过集中流量来改变水体（河流、湖泊、海洋、地下水）的自然流动行为。

2. They can be used to divert, disrupt, or completely stop the natural flow.

本句难点解析：they 指前文的水利结构，be used to do sth. 是被用来去做某事。

本句大意如下：水利结构可以被用来转移、干扰或完全阻止水流的自然流动。

3. Dams (inclusive barrages) are typical water retaining structures that affect closure of the stream and create heading-up afflux.

本句难点解析：这句话的主语是 dam 水坝，that 后进一步说明水坝的作用。

本句大意如下：水坝是典型的蓄水结构，它可以影响水流的截断并产生抬高水位的流动。

4. Water conveying structures are artificial channels cut in the ground and made of either ground materials such as soil and rock (e. g., canals and tunnels) or artificial materials such as concrete and metal (e. g., aqueducts, flumes, siphons, pipelines).

本句难点解析：这句话主语是 water conveying structures，谓语是 are，cut 和 made of 引导的两个短语作为定语并列修饰 channels。

本句大意如下：输水结构是指人工河道，其在地面上开凿，且建造材料为土和岩石这类地面材料（如运河与隧道）或混凝土和金属这类人造材料（如沟渠、引水槽、虹吸管、管道）。

5. Fairly number of alternative layout schemes are compared with respect to technology and economy aspects before the final optimal layout scheme is decided, which should facilitate the construction and the management as well as reduce the investment, on the premise of ensuring project safety.

本句难点解析：这句话的主语是 schemes，谓语是 are，before 引导时间状语从句，which 引导定语从句进一步说明多方面考虑备选方案的好处。

本句大意如下：在最终优化布局方案确定前，相当一部分的备选布局方案会分别在技术和经济方面进行比较，这样能改进施工和管理，并在确保工程安全的前提下减少投资。

6. These mountains and rivers determine the basic features of the state landscape and create superior natural conditions for the two major elements of hydropower resources: runoff and fall.

本句难点解析：这句话的主语是 mountains and rivers，谓语分别是 determine 和 create。

本句大意如下：这些山脉和河流决定了国家自然景观的基本特征，并创造了优越的自然条件，包括水力资源的两大主要元素：径流和落差。

7. following the progress in the construction of large-scale water resources and hydropower pro-

jects, China has reached international level concerning the research, design, and construction technologies.

本句难点解析：following 引导状语从句，这句话的主语是 China，谓语是 has reached。

本句大意如下：伴随着大规模水利项目的建设过程，中国已经在科研、设计和施工技术各方面达到国际领先水平。

8. For this reason, the investigations may comprise a large number of matters, and some or all of them will influence the selection of the project site and scale.

本句难点解析：and 并列两个句子，主语分别为 investigations 和 some or all，谓语分别为 comprise 和 influence。

本句大意如下：基于这个原因，调查的内容可能包含许多因素，这些因素中一部分或者全部将对工程的选址和规模有所影响。

9. Most of these water supply systems originate from the reservoirs of multi-purposed hydraulic projects.

本句难点解析：这句话的主语是 most of these water supply systems，谓语是 originate。

本句大意如下：这些供水系统大多源于多功能水利项目的水库。

10. However, the negative impacts of the hydraulic project on the environment should be never overlooked in the design and construction.

本句难点解析：这句话的主语是 impact，谓语是 be overlooked。

本句大意如下：但是，水利项目对环境的负面影响不应该在设计和施工中被忽略。

Exercises

Translate the following phrases into Chinese.

1. silt mitigation
2. downstream
3. partially submerged in water
4. water retaining structure
5. at the heel
6. temporary diversion structures
7. permanent retaining structures
8. per unit of area
9. water supply reservoir
10. permit infiltration

Translate the following sentences into Chinese.

1. It is common that a number of hydraulic structures (i.e., hydraulic works) of general or special purposes are constructed to form a single or integrated hydraulic project to comprehensively serve foregoing purposes.

2. A gravity dam is a concrete structure resisting the imposed actions by its weight and section without relying on arch.

3. There are two state standards in China for the classification of water resources and hydropower projects with corresponding design safety standards, due to the organization reshuffling history of the state central government.

4. It is a sad fact that thermal power stations, especially those of coal-burned types, discharge a lot of ash and noxious gases into the atmosphere and foul up large territories.

5. A reservoir might significantly make an excellent site for various recreational facilities and health farms. Sometimes, small reservoirs are built specially for recreation purposes.

Part 3

New Technology in Civil Engineering

Unit 11
Industrialized Housing Systems Construction in China

■ 11.1 Introduction to Industrialized Housing Systems

As a typical developing country, China has approximately sixty percent of the population in rural areas. Meeting the rigid demand of rural housing has become a critical issue in Chinese rural development; however, due to the lack of advanced and feasible construction techniques and effective design concepts among Chinese countryside, it always remains a backward state of housing development in these rural areas. The main problems are manifested in several aspects including waste of land resource and construction materials, poor residential functionalities, obvious disorder in construction and building layout, low quality and short lifecycle of house, dissatisfying national building standards, and lack of sustainability. Generally residence construction in Chinese rural areas is organized in disorder. Hence, it's inevitable to find an innovative method for solving rural housing construction problems. Learning from successful construction experience of developed countries, rural housing industrialization helps to achieve good residential functionality and saves construction materials and improves housing quality.[1]

The concept of industrial housing comes from the housing industrialization, which was first proposed in the 1960s by the Japanese MITI. Chinese government first mentioned the concept of housing industrialization in 1994. The housing industrialization is to organize the housing construction and management through social production.[2] It is market-oriented and supported by building material and other light industries. Through housing industrialization, a variety of housing components and parts, finished products, and semi-finished products are produced, transported and assembled on the site. In this way, the entire housing production process from design, component and part production, sales, after-sale services and other various links is integrated into a complete production and management form. Thus, a residential house produced and built in the above mentioned manner is called industrial housing.

Housing industrialization is a concept of producing and managing buildings through massive production approach. In particular, some parts of Chinese rural areas began to realize rural building's life cycle commercialization, standardization, industrialization which enables owners to benefit from saving energy, raw materials and land, improving residents' living conditions and environmental quality, gradually forming the modern village housing industry. Therefore, the transformation of Chinese rural housing mode can be achieved by housing industrialization as a guideline and innovation as well as integrating the design, construction, manufacturing, and supplying process. Different residential areas should consider their own local social economic and technological conditions, also their region-related cultural traditions instead of having a universal construction plan, thus it is necessary to conduct the feasibility analysis of determining whether the region is available with industrialization requirements before starting and implementing industrialized construction. Large gap exists among China's rural areas, so it's also critical to distinguish these areas by analyzing villager's housing consumption tendency, the rural residential housing market strength and national investment to its housing industry, in other words, by evaluating economic, social and cultural factors of these rural areas as to define different levels of rural housing industrialization. According to current research on industrialized construction, the main factor of conducting industrialization is to have a mature standardized building components system, meanwhile enough input for modular building component manufacturing and people's ability and desire to pay for industrialized house are also essential factors, housing demand, construction technology, transportation capacity, supply chain characteristics and national policy support. Construction Method Selection Model (CMSM) is proposed to solve the problem of choosing a specific construction mode for a particular region. CMSM has a strategic level for general environmental feasibility analysis and a tactical level for construction method analysis.[3] With two levels of decision-making process, the characteristics of the local housing market, economic development and local construction regulations decision-

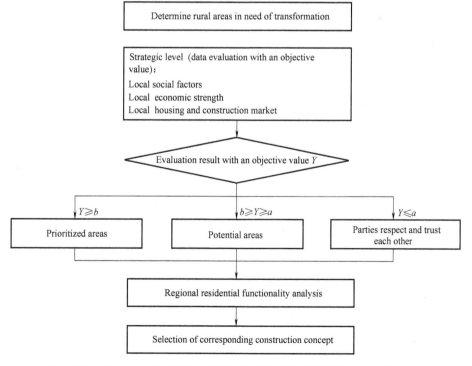

Fig. 11-1 Construction Method Selection Model (CMSM) for rural areas.

making, building quality and performance analysis, building component standardization and construction technologies availability are considered in a comprehensive way. Hence for Chinese rural areas, a CMSM (see Fig. 11-1) will be presented for assessing these housing industrialization related factors as to define the appropriate way of implementing housing industrialization in a particular rural area.

In a word, according to distinctive local conditions, housing industrialization in rural areas should be rationally decided as to adapt to the rural housing market and prolonged building life cycle, also resource coordination and mutual promotion of regional industrial center region should be addressed when considering prerequisites of the decision for developing housing industrialization within each area.

11.2 Types of Industrialized Housing Systems (IHS)

In general, the construction methods can be classified into four categories: (1) conventional method; (2) cast-in situ; (3) composite method; (4) fully prefabricated. The last three construction methods are considered nonconventional and all types are categorized under them. The following paragraphs give an overview about construction methods.

11.2.1 Conventional Construction Method

In the conventional construction method (reinforced concrete frames and brick as infill), beam, column, wall, and roof are cast-in situ using timber formwork while steel reinforcement is fabricated on site.[4] This method of construction is labor intensive and involves three separate trades, namely, steel bending, formwork fabrication, and concreting. Skilled carpenters, plasterers, and brick workers are also involved in this method. The process can be hampered by bad weather and unfavorable site conditions. The application of prefabrication technology to this construction process is a solution to these problems.

11.2.2 Cast-In Situ Construction Method

In this method, lightweight prefabricated formwork that is easily erected and dismantled is used to replace traditional timber formwork.[5] The formwork is a structure—usually temporary but it can be partly or wholly permanent—that is designed to contain fresh fluid concrete, form it into the required shape and dimensions, and support it until it cures sufficiently to become self-supporting. The most commonly used prefabricated formwork materials in the construction are (1) metal products, such as steel and aluminum, (2) plastics products, (3) wood products, (4) ferrocement, (5) fiber products, (6) special timber, and (7) polyethylene formwork.

Those prefabricated formwork materials can be reused many times. It's a fact that repetitive utilization of formwork normally results in cost savings for a project, especially when a large number of housing units is involved. These formworks have clearly shown a construction speed faster than timber formwork.

There are many types of formwork: tunnel form, modular moulding metal form, column and beam form, and permanent form. The tunnel formwork system uses half and completed tunnels. A half tunnel is composed of two panels (vertical and horizontal) set at a right angle and supported by struts and props. The walls and slabs are cast in a single operation. This method leaves no joints, and therefore leakage is unlikely to happen. The modular moulding metal form is also a tunnel formwork system but with only complete walls being cast in situ and precasting slabs or casting in situ slabs, by the conventional methods, are erected onto the completed walls.

11.2.3 Composite Construction Method

The composite construction method, sometimes called partially industrialized, has the objectives of improving quality, reducing cost, and shortening construction time. The concept of "partially by industrialized" derives from the composite nature of full industrialization and conventional construction methods. In this method of construction, certain elements, those that can be standardized, are prefabricated in the factory while others are cast in site. Normally this method would involve the assembly of precast elements, such as floors, slabs, infilled wall, bathrooms, staircases, etc., into place for incorporation into the main unit. Columns and beams are usually cast in situ as these are a relatively easier and less time-consuming part of the operation.

11.2.4 Fully Prefabricated Construction Method

The fully prefabricated construction method can be classified into two main categories. The first is on-site prefabricated and the second is off-site prefabricated (factory producing).[6] On-site precasting consists of casting the floor and roof slabs on top of one another so that they can be lifted into place once the columns and jacking equipment are in place. Cast precast elements at or near the construction site serve to transfer some of the advantages of the factory to the site and in some circumstance proves to be economically and organizationally the most satisfactory method. In the off-site fabrication, some or all components of the building are casting or preparing away from its final position. By transferring the construction operations to the factory, a good quality of product is more easily attainable, and materials and suppliers are much improved and can arrive in economically large loads.

11.3 Selecting Housing Industrialization Grading Indicators

11.3.1 Analysis of Housing Industrialization Factors

The core concept of rural housing industrialization is residential building design standardization and standardized production and installation of building components.[7] Comparing different regions, rural areas with different levels of development have to adapt to the industrialization path of its own needs. The main factors affecting the industrialization path consist of economic factors, transportation and social factors, as Chinese National Comfortable Housing Project has been successfully implemented in some rural areas indicating this housing industrialization is feasible in areas with good economic conditions and convenient transportation. According to United Nations statistics of residential consumption of more than 70 countries, per capita GNP and the housing industry are positively correlated, and the housing industry enters the rapid development stage when per capita GDP rises to about \$ 800, thus a solid economic condition is a critical prerequisite for industrialization. Light steel residence building is more expensive than the traditional rural house built with bricks and wood, and whether this rural area is affordable to develop industrialized building with sustainable materials or villagers are capable to buy these houses without breaking housing contract is important for deciding adoption of rural housing industrialization. Meanwhile considering most of the villages in China are very underdeveloped, the housing industrialization investment must rely partially on government support and national policy such as National Urbanization Project. Besides, industrialized residential construction requires also the need for effective coordination between building components' production, transportation and assembly. Since most rural population scattered in the mountains or along rivers, the terrain is more complex than urban areas. The factor concerning transportation of standardized building component is more critical, and a set of sound transportation infrastructure and transportation network is beneficial for assisting building components transporting logistics. From social perspective, dissemination capabilities of new ideas is relatively low in China rural areas, rural residents hardly have the perception of the housing industrialization, instead they are over-relying on the construction of brick houses and they do not tend to use new construction materials and new building technology, thereby it is necessary to increase the awareness of energy saving sustainability of rural residents when promoting concept of rural housing industrialization.

11.3.2 Determining housing industrialization grading indicators

When selecting evaluation indicators, not only the strong independence between indicators should be considered reflecting the basic characteristics of each factor, but also there should pay attention to each indicator's operability and measurability for the transverse and longitudinal comparisons. Based on economic factor, social factor and transportation convenience along with the statistical targets in Yearbook of China's Rural Residents Statistics, rural housing industrialization grading indicators are selected accordingly (see Table 11-1). They are 8 second class indicators, and note that five indicators are processed based on the original indicators stated in Yearbook of China's Rural Residents Statistics which are presented as follows:

1) Housing consumption tendency = yearly living expenses/yearly income.

2) House purchase indemnificatory = yearly living expenses/value of newly bought house in a year.

3) Housing investment driven ratio = newly built house space/housing investment.

4) Labor education level = ratio of resident with a junior college degree in a rural family.

5) Degree of dissemination of non-brick building construction = concrete building floor areas/brick (or wood) building floor areas.

Table 11-1 Rural housing industrialization grading indicators

1st class indicator	2nd class indicator
Economical element	Housing consumption tendency of rural residents House purchase indemnificatory of rural residents Fixed real estate investment of rural residents Consumption level of rural residents Housing investment driven ratio in rural areas
Transportation Social element	Transportation, warehousing and postal assets investment Personal education level of a rural family Degree of dissemination of non-brick building construction

■ 11.4 Basic Methodology of Hierarchical Clustering Method

11.4.1 Hierarchical Clustering Method

Hierarchical clustering is a method used to determine clusters of similar data points in multi-dimensional spaces. There are a number of methods of determining the distances between clusters, such as single link, average link, complete link of graph method and centroid, median and minimum variance of geometric method (Olson 1995). This paper is implementing hierarchical analysis using MATLAB since this software has strong data processing, data analysis and dynamic simulation capabilities, which is efficient for clustering analysis of masses data. Hierarchical clustering method based on MATLAB data analysis includes the following steps:

1) Data normalization of which the formula is as below:

$$x_{ij}^* = \frac{x_{ij}-x_i}{s_i}, (i=1,2,\cdots,m; j=1,2,\cdots,n) \tag{11-1}$$

2) Calculating the distances between the clusters with standardized Euclidean distance. Their standardized Euclidean distance is calculated in the formula as below of which:

$$\boldsymbol{x}_i = (x_{1i}, x_{2i}, \cdots, x_{mi})^T, \boldsymbol{x}_j = (x_{1j}, x_{2j}, \cdots, x_{mj})^T$$

$$d(x_i, x_j) = \frac{1}{s_i}\sqrt{\sum_{k=1}^{m}(x_{ki}-x_{kj})^2} \tag{11-2}$$

3) Clustering points according to their distances. Based on variance analysis, MATLAB is automatically applying the Ward method (deviation square method) so that the individuals within the same classes have minimized squared deviations, and squared deviations between classes are maximized.

11.4.2 Clustering Analysis of Grading Housing Industrialization

The original data is collected from *China Rural Statistical Yearbook* 2012, containing information of nationwide rural residents' yearly income, living expenses, consumption structure, fixed asset investment, newly built housing space, etc. Since a part of China rural areas situate in extreme climatic regions, such as Tibet with big temperature difference within a day, these areas are not suitable for implementing housing industrialization which does not meet special building standards of heat insulation. Meanwhile considering several rural areas have strict cultural adherence, it is more rational to conserve the minority culture and specific features of traditional villages instead of developing industrialization. Hence, this paper aims to analyze 23 rural areas nationwide except Tibet, Xinjiang, Neimenggu, Guangxi, Yunnan, Ningxia, Qinghai, Gansu, Hongkong, Macau and Taiwan Province. According to preliminary analysis of raw data, it is evident that these areas are different from each other by certain characteristics; descriptive statistics including maximum, minimum, average value and standard deviation are shown in Table 11-2. By executing MATLAB code of hierarchical clustering with Linkage function calculating the binary data of hierarchy tree, after the maximum number of iterations is reached, the category polymerization performs to a converging point and the distance between each of the two provinces is illustrated in distribution graph, namely systems Dendrogram (see Fig. 11-2). The number of final clustering category is 4, and the first cluster includes provinces of Shanxi, Heilongjiang, Jiangxi, Hunan, Hubei, Jilin and Sichuan (represented by D, G, M, Q, P, F, U respectively). The second cluster includes provinces of Hebei, Anhui, Shaanxi, Chongqing, Hainan and Henan (represented by C, K, W, T, S, O respectively). The third cluster includes provinces of Liaoning, Guangdong, Shandong, Fujian and Guizhou (represented by E, R, N, L, V respectively). The fourth cluster includes provinces of Beijing, Zhejiang, Tianjin, Jiangsu and Shanghai (represented by A, J, B, I, H respectively). Meanwhile the Cophenet function is used to verify the effectiveness of clustering results of which the value is more approximate to 1.00 indicating better clustering reliability. With an output Cophenet value of 0.9068, it can be concluded that the results are reliable and all the 4 clusters are significantly different from each other.

Table 11-2 Descriptive statistics of rural areas grading indicators

Rural housing industrialization grading indicators	Maximum value	Minimum value	Average Value	Standard deviation
Housing consumption tendency of rural residents	0.22	0.07	0.13	0.03
House purchase indemnificatory of rural residents	1.84	0.38	1.12	0.42
Fixed real estate investment of rural residents	745.00	2.60	246.40	191.51
Consumption level of rural residents	18512.00	4448.00	8325.78	3532.09
Housing investment driven ration in rural areas	18512.0	4448.00	8325.78	3532.09
Transportation, warehousing and postal assets investment	66.20	0.00	20.91	15.85
Personal education level of a rural family	13.50	1.20	3.62	2.67
Degree of dissemination of non-brick building construction	77.00	0.44	7.20	15.68

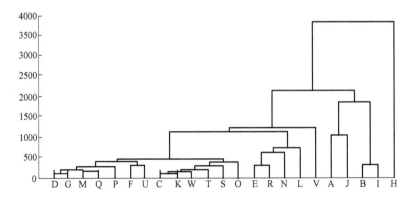

Fig. 11-2 Clustering dendrogram.

11.5 Region-Based Housing Industrialization Grading Analysis

11.5.1 Prioritized, Potential, Support-required Areas

1) Prioritized areas for rural housing industrialization.

The prioritized areas represented by the fourth cluster are pioneer areas suitable for housing industry.[8] The rural residents in Shanghai, Zhejiang and other areas with higher income levels and housing consumption tendency will to pay more on housing as to improve the quality of living conditions. This type of region usually has sound financial foundation and high housing credit which is prioritized as a leading role for housing industrialization by National Demonstration Projects. Through urban-rural industrialization integration strategy, the establishment of rural industrialization development group will exert a positive influence in improving the level of industrialization radiation and the integrity of industrialized building component supply chain, hence reducing the cost of building industrialized buildings.

2) Potential areas for rural housing industrialization.

The potential areas represented by the third cluster are mostly concentrated in the east part of China. Although rural residents in these areas comparatively have substantial income, there is less guarantee for performing the housing contract wells, for instance, the values of house purchase indemnificatory in Guangdong, Liaoning, Fujian are all less than the average level of 1.12, while the possible reason for this gap is that the rural residents lack the needs for newly built buildings. Therefore, these areas should focus on improving rural housing functionality, such as making full use of methane and waste water and enhancing the energy efficiency and insulation efficiency of buildings in northeastern region as to improve the overall living standards. At the same time the government should provide adequate investment to rural housing construction and industrial market development, rural areas with high population density may rely on the urban housing industrialization center through setting up material distribution center, the development of these potential areas will be the reserve forces of nationwide rural housing industrialization.

3) Support-required areas lack of industrialized building prerequisites.

The support-required areas are represented by the first and second clusters, whose rural residents have the lowest levels of yearly income and consuming power, for instance, Chongqing rural resident's consumption tendency and house purchase indemnificatory values are minimum. Heilongjiang registered the lowest value in personal education level of a rural family and degree of dissemination of non-brick building construction. The most significant feature of these areas is the poor economic condition and it is not beneficial to promote rural housing industrialization compulsively. However, if the building cost is minimized in these areas, the housing industrialization can be possibly available in the way of absorbing spillover advantages and obtaining indirect support of adjacent developed regions. And government should still encourage rural residents and enterprises in these areas to gradually accumulate the initial capital of housing industrialization, aiming on semi housing industrialization.

11.5.2 Patterns for Region-Based Housing Industrialization

The rural housing industrialization level is significantly different from one region to another, and obviously it is easier to promote industrialized houses in prioritized areas with good economic conditions and comprehensive construction technologies. According to the clustering results, relative rural housing construction modes can be also divided into three patterns (see Fig. 11-3).

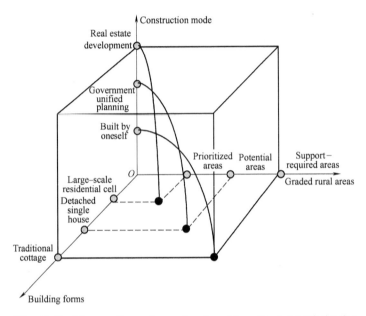

Fig. 11-3 Three patterns for region-based housing industrialization.

The first pattern relates to the prioritized areas. Since they are close to comprehensive industrialized building component distributing system, it is possible to form large-scale industrialized building cells. The second pattern focuses on potential areas usually featured by government unified township planning. These areas are suitable for building detached single house for each rural family which meets rural resident's basic agricultural production and living requirements. The third pattern is in line with the support-required areas of which the houses are lack of the scientific design wasting a lot of construction materials. Villagers tend to build their own cottage with poor security and building performance, hence it is necessary to propagate the core concept of rural housing industrialization among these rural residents and government should strengthen the recognition and usage of standardized design and construction dendrogram. In supported areas, rural cottages with poor performance should be replaced gradually by detached single house of potential areas, which shows the curve of support-required areas is approaching to the curve of potential areas. Among these three patterns, each curve of a specific housing industrialization grading category is approaching from low level to high level so that the disorder in rural housing construction will be gradually eliminated.

11.6 Development Bottlenecks of Housing Industrialization in China

1) The national standard system for industrial housing has not been issued yet so far in China. It is difficult to carry out large-scale precast production. Moreover, a detailed technical specification to guide the prefabricated assembly and construction is still in the preparation stage in China. The housing quality cannot be guaranteed.

2) The entire housing production process includes design, component production, sales, after-sale services and in-site assembly and construction. In China, there are still no super enterprises to perform the entire production process by alone. Also there is no multi-body coordination mechanism formed and implemented. The discrepancy in opinions from different participants has significantly increased transaction cost.

3) The management mode is a fundamental change not only in construction methods but also in other aspects like project bidding, project valuation model, prefabricated parts production, quality control and final acceptance. A construction management mode applied to industrial housing has not been established yet. As the traditional construction management mode cannot be applied to the industrial housing production and construction.

4) Unlike the traditional labor-intensive construction mode, the industrial housing requires professional and technical personnel at the construction site. There is a large lack of technical staff who know how to design and construction industrial houses in China.

11.7 Advantages and Disadvantages of Industrialized Housing Systems

The quality, speed of construction, and cost saving are the main emphases given in the housing construction industry in China.[9] The savings in labor cost and the savings in material cost are also the major advantages of the Chinese Industrial Housing Systems (IHSs). The control in using materials, such as steel, sand, and timber, will result in substantial savings on the overall cost of the project.

The IHS construction activities are highly capital intensive. This is the main disadvantage of the IHS. The heavily mechanized approach has displaced a substantial number of the labor force from the building construction industry. In some IHSs there is a tremendous need for expert labor at the construction site. Therefore, extra costs are needed to train the semiskilled labor force for highly skilled jobs.

The main reasons for delay in early completion of projects in IHS construction industry are supply delay, bad weather, and shortage of raw material. In some cases, the main reason for the delay was the lack of labor experience. This is because certain types of IHS construction are still new in China and the labor force is still not familiar with the special erection procedure required by those systems.

Almost all the IHSs in China are suitable for all number of stories, especially for three to five-story buildings. At the same time, all the IHSs in China are very much suitable for all classes of construction from the unit cost point of view, which are arranged from low-cost house class to high-cost house class.

11.8　Key Implementations to Housing Industrialization

Different housing industrialization implementation measures should be taken according to the three levels of rural along with analyzing the key industrialization procedures and qualifications in these prioritized, potential and support-required areas.[10] First, it is essential to formulate the production specifications for the standardized building components and unified modules system based on the area's building construction features and geographic characteristics. Then the housing functionality needs be improved by professional and technical identification of each specific building performance indicators such as building fire protection, a seismic design, and insulation level. Also the technical specification standards should be proposed for industrialized design of reinforced concrete buildings, especially for those in extreme weather conditions (such as the pretty cold China Northeast regions and southwest regions with earthquake-prone risks). From government perspective, rural housing industrialization needs government policy protection and financial support in resolving the current problem of Chinese rural homestead system and sponsoring housing subsidies. Government should also leveraging small and medium enterprises in rural areas in providing construction materials and building design who will also help foster housing industrialization market, besides, the establishment of a regional distribution center will perform as a source sharing node in suburban areas, reducing the cost of transporting of industrialized building components. Implementing rural house industrialization is a long-term project, which requires the participation of technologists, entrepreneurs, financial sectors and national political sectors. The implementation measure should be constantly revised and adjust the key housing industrialization process to the right path in order to find a more rational way of implementation.

New Words and Expressions

feasibility　*n.*　可行性，可能性
prerequisite　*n.*　先决条件
lightweight　*adj.*　轻质的
dismantle　*v.*　拆卸，拆除
sustainability　*n.*　持续性，永续性；能维持性
residents　*adj.*　居住的；*n.*　居民
dissemination　*n.*　宣传；散播
dendrogram　*n.*　树状图

 Notes

1. Learning from successful construction experience of developed countries, rural housing industrialization helps to achieve good residential functionality and saves construction materials and improve housing quality.

本句难点解析：learning 引导的是状语从句，这句话的主语是 rural housing industrialization，谓语是 helps 和 saves。

本句大意如下：学习发达国家成功的建造经验，乡村住宅产业化有助于实现更好的住宅功能，节省建造材料，改善房屋质量。

2. The housing industrialization is to organize the housing construction and management through social production.

本句难点解析：这句话的主语是 housing industrialization，谓语是 is to。

本句大意如下：住宅产业化就是通过社会化生产来组织房屋施工和管理。

3. CMSM has a strategic level for general environmental feasibility analysis and a tactical level for construction method analysis.

本句大意如下：建造方法选择模型可以提供战略层面的环境可行性分析以及战术层面的建造方法分析。

4. In the conventional construction method (reinforced concrete frames and brick as infill), beam, column, wall, and roof are cast-in situ using timber formwork while steel reinforcement is fabricated on site.

本句难点解析：这句话的主语是 beam, column, wall and roof，谓语是 are，while 引导状语从句。

本句大意如下：传统的建造方法中，梁、柱、墙和楼板均用木模板现浇，钢筋在现场焊接加工。

5. In this method, lightweight prefabricated formwork that is easily erected and dismantled is used to replace traditional timber formwork.

本句难点解析：这句话的主语是 formwork，谓语是 is used to，that 引导定语从句修饰 formwork。

本句大意如下：这种方法采用容易安装和拆除的轻型预制模板来取代传统的木模板。

6. The first is on-site prefabricated and the second is off-site prefabricated (factory producing).

本句难点解析：这句话主语是 the first 和 the second，谓语均为 is。

本句大意如下：第一种是现场预制，第二种是工厂预制。

7. The core concept of rural housing industrialization is residential building design standardization and standardized production and installation of building components.

本句难点解析：这句话的主语是 core concept，谓语是 is，宾语为两个并列的标准化。

本句大意如下：农村住宅产业化的核心概念是住宅建筑的设计标准化，以及构件生产和安装的标准化。

8. The prioritized areas represented by the fourth cluster are pioneer areas suitable for housing

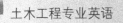

industry.

本句难点解析：这句话的主语是 areas，represented by the fourth cluster 作为定语修饰主语，谓语是 are。

本句大意如下：以第四集群为代表的优先区域是适合优先发展住宅产业化的地区。

9. The quality, speed of construction, and cost saving are the main emphases given in the housing construction industry in China.

本句难点解析：given 为定语从句修饰 emphases，主语是 quality，speed 和 cost saving，谓语为 are。

本句大意如下：质量、施工速度和成本是住宅建筑产业化在中国的主要关注点。

10. Different housing industrialization implementation measures should be taken according to the three levels of rural along with analyzing the key industrialization procedures and qualifications in these prioritized, potential and support-required areas.

本句难点解析：这句话的主语是 measures，谓语是 be taken，along with 作定语修饰 levels of rural。

本句大意如下：根据乡村的3个等级，并分析在这些有潜力且优先需要支持的地区中产业化的核心步骤和条件，针对不同结果采用不同的住宅产业化措施。

Translate the following phrases into Chinese.

1. housing industrialization

2. industrial housing

3. consumption tendency

4. critical prerequisite for industrialization

5. multi-dimensional spaces

6. clustering analysis

7. urban-rural industrialization integration strategy

8. overall cost of the project

9. sponsoring housing subsidies

10. a rational way of implementation

Translate the following sentences into Chinese.

1. Meeting the rigid demand of rural housing has become a critical issue in Chinese rural development; however, due to the lack of advanced and feasible construction techniques and effective design concepts among Chinese countryside, it always remains a backward state of housing development in these rural areas.

2. It's a fact that repetitive utilization of formwork normally results in cost savings for a project, especially when a large number of housing units is involved.

3. Hierarchical clustering is a method used to determine clusters of similar data points in multi-dimensional spaces.

4. The management mode is a fundamental change not only in construction methods but also in other aspects like project bidding, project valuation model, prefabricated parts production, quality control and final acceptance.

5. From government perspective, rural housing industrialization needs government policy protection and financial support in resolving the current problem of Chinese rural homestead system and sponsoring housing subsidies.

Unit 12

Passive Base Isolation with Merits and Demerits Analysis

■ 12.1 Introduction

The conventional approach to seismic-resistant design is to incorporate adequate strength, stiffness and inelastic deformation capacity into the building structure so that it can withstand induced inertia forces. This was with the presumption that during strong ground motion, whenever inertia forces exceed their design earthquake levels, the structure will dissipate this excess energy through deformations at predefined locations scattered over the structural framework. It was observed that, even with members designed for ductility, the structures did not always perform as desired, which could be because of reasons such as:

1) Strong-column weak-beam mechanism failed to develop as a result of stiffening effect of walls being present.

2) Creation of short columns because of changes in wall layout, introduced later.

3) Poor concreting at joints due to reinforcement congestion.

On experiencing failures during earthquakes, it was realized that a design based only on the principle of incorporating ductility as a safeguard against seismic effects needs a critical review. In their search for alternative design strategies to minimize magnitude of inertia forces, engineers came up with the innovative idea of introducing a flexible medium between supporting ground and the building, thereby decoupling the structure from energy-rich components of seismic ground motion. This strategy came to be known as the base isolation method.

The frequency of vibration of low-to-medium-rise buildings falls in the range where earthquake energy is high. This has often resulted in considerable damage to such buildings due to the enormous destructive strength of an earthquake. The prevailing earthquake-resistant design methods tacitly accept that during a major earthquake, structural damage, sometimes substantial in magnitude, is unavoidable. However, the extremely high cost of repairing and rebuilding damaged structures motivated designers to re-look at the concept of incorporating a flexible medium between ground and the building, which was first attempted in its rudimentary form, over a century earlier.

The concept of seismic isolation has become a practical reality within the last 20 years with the development of multilayer elastomeric bearings, which are made by vulcanization bonding of sheets of rubber to thin steel reinforcing plates. These bearings are very stiff in the vertical direction and can carry the vertical load of the building but are very flexible horizontally, thereby enabling the building to move laterally under strong ground motion. Their development was an extension of the use of elastomeric bridge bearings and bearings for the vibration isolation of buildings. In recent years other systems have been developed that are modifications of the sliding approach. The concept of base isolation is now widely accepted in earthquake-prone regions of the world for protecting important structures from strong ground motion, and there are now many examples in the United States and Japan. A smaller number of base-isolated buildings have been built in New Zealand and in Italy, mainly for large and important buildings. Demonstration projects that apply low-cost base isolation systems for public housing in developing countries have been completed in Chile, the People's Republic of China, Indonesia, and Armenia.

It is not surprising that most applications are for important buildings that house sensitive internal equipments. The basic dilemma facing a structural engineer charged with providing superior seismic resistance of a building is how to minimize interstory drifts and floor accelerations. Large interstory drifts cause damage to nonstructural components and to equipments that interconnects stories. Interstory drifts can be minimized by stiffening the structure, but this leads to amplification of the ground motion, which leads to high floor accelerations, which can damage sensitive internal equipments. Floor accelerations can be reduced by making the system more flexible, but this leads to large interstory drifts. The only practical way of reducing simultaneously interstory drifts and floor accelerations is to use base isolation; the isolation system provides the necessary flexibility, with the displacements concentrated at the isolation level.

The concept of base isolation is quite simple. The system decouples the building or structure from the horizontal components of the ground motion by interposing structural elements with low horizontal stiffness between the structure and the foundation.[1] This gives the structure a fundamental frequency that is much lower than both its fixed-base frequency and the predominant frequencies of the ground motion. The first dynamic mode of the isolated structure involves deformation only in the isolation system, the structure above being to all intents and purposes rigid. The higher modes that produce deformation in the structure are orthogonal to the first mode and, consequently, to the ground motion. These higher modes do not participate in the motion, so that the high energy in the ground motion at these higher frequencies cannot be transmitted into the structure. The isolation system does not absorb the earthquake energy, but rather deflects it through the dynamics of the system; this effect does not depend on damping, but a certain level of damping is beneficial to suppress possible resonance at the isolation frequency.

The first use of a rubber isolation system to protect a structure from earthquakes was in 1969 for an elementary school in Skopje, Yugoslavia. Later, a building in Sebastopol in Crimea was supported on steel bearings. It is reported to have performed satisfactorily during an earthquake in 1977. Such satisfactory performance of this and other initial attempts established that base isolation is a viable, and in some cases a superior option, to conventional design methods. This prompted scientists and equipment manufacturers to put in considerable research and development efforts that brought about improvements and practical solutions in the last quarter of the twentieth century.

Rapid development of this technology saw the introduction of newer and newer rubber products that possessed the desired elasto-plastic characteristics (i. e. high stiffness at low strains and low stiffness at high strains) which propelled base isolation to become an internationally accepted method of reducing earthquake-induced inertia forces. Availability of high-speed computing capability supported by reliable software to efficiently tackle complex design issues complemented by large shake table testing facilities played a major contributory role in base isolation technology gaining a firm foothold.

The concept of base isolation has also provided a rich source of theoretical work, both in the dynamics of the isolated structural system and in the mechanics of the isolators themselves. This theoretical work, widely published in structural engineering and earthquake engineering journals, has led to design guidelines for isolated structures and design rules for isolators. In parallel in the 1980s, efforts got under way to develop codes specifically applicable to seismically isolated buildings and in due course, code provisions were introduced commencing with the publication in 1986 of "Tentative Seismic Isolation Design Requirements" by Seismology Committee of Structural Engineers Association of California (SEAOC). These guidelines and their subsequent revisions proved to be the precursor to the introduction of design guidelines by Federal Emergency Management Agency (FEMA) followed by the "Uniform Building Code" and "International Building Code". In all these documents, the underlying design philosophy has been life safety under major earthquake and acceptable performance of the building and its constituents under design earthquake.

■ 12.2 Concept of Base Isolation

To diminish vulnerability of a building to damage during an earthquake, base isolation has emerged as a viable structural option. It is a sophisticated practical solution to improve seismic response of a building by minimizing the structural damage, which was earlier taken to be unavoidable during strong ground motion. Because of the low horizontal stiffness of this deformable medium, it alters the fundamental period of a stiff structure such that it is significantly higher than that of the high energy imparting ground motions. As a result, for its fundamental mode of vibration, the superstructure is subjected to much lower inertia forces with consequent reduction (Fig. 12-1) in base shear.

On the flip side, if the founding stratum is soft, then there is a distinct possibility of the enhanced period due to base isolation being close to the period where an earthquake is likely to have considerable energy. Such a situation can lead to an increase in the response. Thus, it can be said that base isolation is best suited for buildings with a high natural fundamental frequency (T less than about 1s) and preferably those supported on rock or stiff soils. Reinforced concrete moment framed buildings up to about 8~9 storeys and those with shear walls up to 12~15 storeys are said to be ideal candidates for base isolation.

隔震的概念

Unit 12 Passive Base Isolation with Merits and Demerits Analysis

If a framed structure were supported on hard strata, it would deform as in Fig. 12-2a. However, when supported on isolators, the lateral displacement in the first mode is concentrated at isolator level while the superstructure behaves almost like a rigid body. As a result, for buildings supported on base isolators, there is less need to provide ductile energy dissipating regions (e. g. near beam-column joints) as in conventional fixed based structures. However, in view of limited experience of such systems and as a matter of abundant precaution, codes recommend retention of the present form of ductile detailing. This is also to obviate possible brittle failure under a maximum credible earthquake (MCE) in the region or earthquakes with long period energy inputs. In addition, codes also call for rigorous testing of the proposed isolators and peer review of the design, since it is an emerging technology and failure of an isolator system could prove catastrophic.

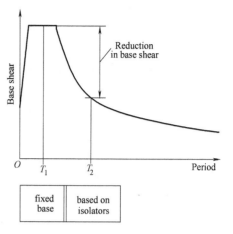

Fig. 12-1 Effect of increased period
(representative diagram).

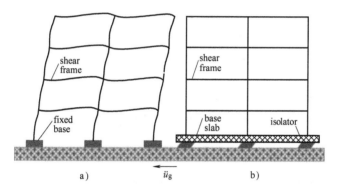

Fig. 12-2 Displaced MDOF system.
a) Fixed bas b) Based on isolators

Isolators should be located such that there is ample access to them for maintenance, repair and replacement, if necessary. A full diaphragm should be employed to distribute lateral loads as uniformly as possible to the isolators. Generally, isolators can be placed at the bottom of columns at basement slab level. This location has the advantage that no special treatment is required for elevator or service lines as they traverse the bearing level. On the other hand, if they are placed at top of basement, then elevator shafts and internal staircases and cladding details may require special treatment below first floor level.

Isolators do substantially enhance the fundamental period of a stiff structure but at the expense of increased building displacements.[2] The first mode deformation occurs at the isolation level only. The real challenge while designing a base isolated structure is the need to control displacements during a major earthquake while maintaining good performance for moderate level earthquakes. Since the superstructure will function essentially as linearly elastic, the structural framework is expected to remain undamaged even during a moderate-level earthquake.

12.3 Base Isolation Systems

Base isolation is now a mature technology and is used in many countries, and there are a number of acceptable isolation systems, the construction of which is well understood. Nevertheless, the concept appears to have an irresistible attraction to inventors, and many new and different systems of isolators are proposed and patented each year. Many of these new systems will prove to be impractical and some might actually be lethal, but the number continues to increase year by year.

Most systems used today incorporate either elastomeric bearings, with the elastomer being either natural rubber or neoprene, or sliding bearings, with the sliding surface being Teflon and stainless steel (although other sliding surfaces have been used). Systems that combine elastomeric bearings and sliding bearings have also been proposed and implemented.

12.3.1 Elastomeric-based Systems

Natural rubber bearings were first used for the earthquake protection of buildings in 1969 for the Pestalozzi School in Skopje, Macedonia. The bearings are large rubber blocks without the steel reinforcing plates used today and compress by about 25% under the weight of the building. The bearings have a vertical stiffness that is only a few times the horizontal stiffness and the rubber is relatively undamped.[3] This system was tested on the shake table at the EERC in 1982. Characteristic of isolation systems of this kind, the horizontal motion is strongly coupled to a rocking motion, so that purely horizontal ground motion induces vertical accelerations in the rocking mode. The system also has foam-glass blocks on either side of a rubber bearing that are intended to act as fuses to prevent movement in the building under wind, internal foot traffic, or low seismic input. The system is still in place and is monitored from time to time.

隔震系统

Since this building was completed, many other buildings have been built on natural rubber bearings but with internal steel reinforcing plates that reduce the lateral bulging of the bearings and increase the vertical stiffness. The internal steel plates, referred to as shims, provide a vertical stiffness that is several hundred times the horizontal stiffness. These multilayered elastomer bearings provide vibration isolation for apartment blocks, hospitals, and concert halls built over subway lines or mainline railroads. In 1975 Derham et al. suggested that this approach could be used to protect buildings from earthquake ground motion, and an intensive experimental and theoretical research program was begun at the EERC to develop this concept. Laminated elastomeric bearings can be differentiated into low-damping or high-damping types.

1. Low-damping natural and synthetic rubber bearings

Low-damping natural rubber bearings and synthetic rubber bearings have been widely used in Japan in conjunction with supplementary damping devices, such as viscous dampers, steel bars, lead bars, frictional devices, and so on. The elastomer used in Japan comprises natural rubber, while in France neoprene has been used in several projects. The isolators have two thick steel endplates and many thin steel shims, as shown in Fig. 12-3. The rubber is vulcanized and bonded to the steel in a single operation under heat and pressure in a mold. The steel shims prevent bulging of the rubber and provide a high vertical stiffness but have no effect on the horizontal stiffness, which is controlled by the low shear modulus of the elastomer. The material behavior in shear is quite linear up to shear strains above 100%, with the damping in the range of 2% ~ 3% of critical. The material is not subject to creep, and the long-term stability of the modulus is good.

Fig. 12-3 Low-damping natural rubber bearing.

The advantages of the low-damping elastomeric laminated bearings are many: They are simple to manufacture (the compounding and bonding process to steel is well understood), easy to model, and their mechanical response is unaffected by rate, temperature, history, or aging.[4] The single disadvantage is that a supplementary damping system is generally needed. These supplementary systems require elaborate connections and, in the case of metallic dampers, are prone to low-cycle fatigue.

2. Lead-plug bearings

The lead-plug bearing was invented in New Zealand in 1975 and has been used extensively in New Zealand, Japan, and the United States. Lead-plug bearings are laminated rubber bearings similar to low-damping rubber bearings but contain one or more lead plugs that are inserted into holes[5], as shown in Fig. 12-4. The steel plates in the bearing force the lead plug to deform in shear. The lead in the bearing deforms physically at a flow stress of around 10MPa (about 1500psi), providing the bearing with a bilinear response. The lead must fit tightly in the elastomeric bearing, and this is achieved by making the lead plug slightly larger than the hole and forcing it in. Because the effective stiffness and effective damping of the lead-plug bearing is dependent on the displacement, it is important to state the displacement at which a specific damping value is required. Lead-plug bearings have been extensively tested in New Zealand, and there are very complete guidelines on their design and modeling. Buildings isolated with these bearings performed well during the 1994 Northridge and 1995 Kobe earthquakes.

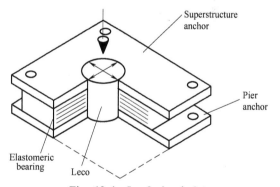

Fig. 12-4 Lead-plug isolator.

3. High-damping natural rubber systems (HDNR)

The development of a natural rubber compound with enough inherent damping to eliminate the need for supplementary damping elements was achieved in 1982 by the Malaysian Rubber Producers' Research Association (MRPRA) of the United Kingdom. The damping is increased by adding extrafine carbon block, oils or resins, and other proprietary fillers.[6] The damping is increased to levels between 10% and 20% at 100% shear strains, with the lower levels corresponding to low hardness (50~55 durometer) and a shear modulus around 0.34MPa (about 50psi) and the high levels to high hardness (70~75 durometer) and a high shear modulus around 11.40MPa (about 200psi). The methods of vulcanization, bonding, and construction of the isolators are unchanged.

The material is nonlinear at shear strains less than 20% (Fig. 12-5) and is characterized by higher stiffness and damping, which tends to minimize response under wind load and low-level seismic load. Over the range of 20% ~ 120% shear strain, the modulus is low and constant. At large strains the modulus increases due to a strain crystallization process in the rubber that is accompanied by an increase in the energy dissipation. This increase in stiffness and damping at large strains can be exploited to produce a system that is stiff for small input, is fairly linear and flexible at design level input, and can limit displacements under unanticipated input levels that exceed design levels.

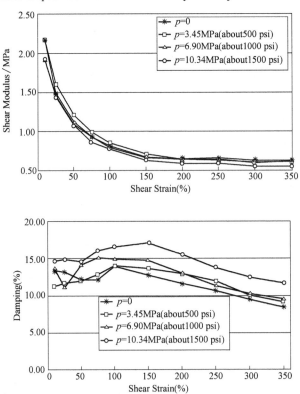

Fig. 12-5 Stress-strain characteristics of high-damping natural rubber isolators.

The damping in the isolators is neither viscous nor hysteretic, but somewhat in between. In a purely linear viscous element the energy dissipation is quadratic in the displacement; in a hysteretic system it tends to be linear in displacement. Tests on a large number of different rubber isolators at the EERC (Fig. 12-6) demonstrate that the energy dissipated per cycle is proportional to the displacement around the value of power 1.5. This characteristic can be exploited so that it is possible to model the bearing response, which combines linear viscous and elastic-plastic elements.

Another serendipitous advantage of the high-damping rubber system is that it provides a degree of ambient vibration reduction. The isolators will act to filter out high-frequency vertical vibrations caused by traffic or adjacent underground railways. This effect was demonstrated in a shake table test program carried out at EERC in 1985.

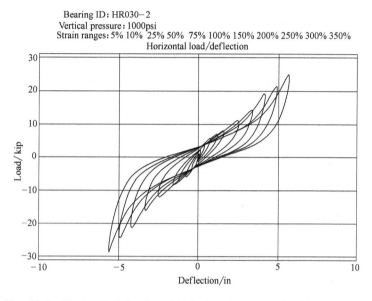

Fig. 12-6 Hysteretic behavior of high-damping natural rubber isolators.

12.3.2 Isolation Systems Based on Sliding

A purely sliding system is the earliest and simplest isolation system to be proposed. A system using pure sliding was proposed in 1909 by Johannes Avetican Calantarients, a medical doctor in England. He suggested separating the structure from the foundation by a layer of talc (shaded portion in Fig. 12-7). As is evident in his diagrams, Calantarients clearly understood that the isolation system reduced accelerations in the isolated building at the expense of large relative displacements between the building and the foundation, for he designed a set of ingenious connections for utilities—in those days restricted to gas lines and sewage pipes—to accommodate these displacements. In fact, Calantarients's system incorporated all the elements now considered necessary in a base isolation system: a method of decoupling the building and the foundation, a method whereby utility lines can withstand large relative displacements, and a wind restraint system.[7]

When sliding isolators are provided, the building is supported on surfaces of stainless steel sliding against a very low friction material like teflon. This permits transmission of shear forces across isolator interface only to the extent of frictional resistance between sliding layers. Such a system is relatively less expensive and is ideal for retrofitting. It is effective over a wide range of frequencies normally prevalent in input ground motions. Another advantage is that the maximum trans-

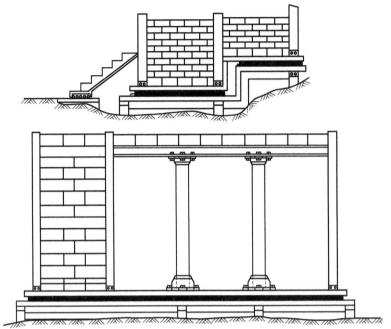

Fig. 12-7 Calantarients's base isolation system using a layer of talc as the isolating medium.

missibility of acceleration to the superstructure is limited to the maximum force that can generate at the frictional interface.

In a sliding isolator, since the frictional force generated is proportional to the weight it supports, the center of mass and center of resistance naturally coincide.[8] As a result, these systems are less sensitive to torsional coupling. Many variations of this fundamental system have been developed, which include a sliding surface with varying curvature, two cylindrical sliding surfaces with different curvatures which are placed perpendicular to each other, etc. However, the system most frequently adopted and described herein is an isolator with a single spherical sliding surface. A schematic of such an isolator is shown in Fig. 12-8.

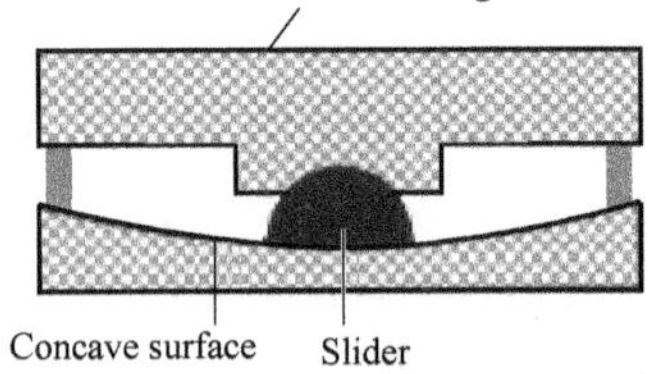

A considerable amount of theoretical analysis has been done on the dynamics of structures on sliding systems subjected to harmonic input or to earthquake input. For example, as a representation of a base-isolated building, Westermo and Udwadia studied the periodic response of a linear oscillator on a coulomb friction sliding interface. Contrary to the

Fig. 12-8 Typical sliding oscillator.

general perception that friction will always reduce the response, they found that the response may be larger than that for the same fixed-base model and that the single degree of freedom model had subharmonic resonance frequencies generated by the sliding interface. The response of a similar model to earthquake input was studied by Mostaghel et al.

The assumption of coulomb friction is generally used in these theoretical analyses but is unlikely to be an accurate representation of real behavior. The most commonly used materials for sliding bearings are unfilled or filled polytetrafluoroethylene (PTFE, or Teflon) on stainless steel, and the frictional characteristics of this system are dependent on temperature, velocity of interface motion, degree of wear, and cleanliness of the surface. Much testing work has been done on these aspects of the mechanical behavior of such sliding components and an extensive review was done by Campbell and Kong.

1. Electricité-de-France system

This system was developed in the early 1970s for application to nuclear power plant facilities. The utility developed a standard nuclear power plant with the safety grade equipment qualified for 0.2g acceleration. When the standard plant was to be located at sites of higher seismicity, it was isolated to keep the equipment acceleration levels below the qualification value. The system combines laminated neoprene bearings (essentially standard bridge bearings manufactured to higher quality control standards) with lead-bronze alloy in contact with stainless steel, the sliding surface being mounted on top of the elastomeric bearing. The coefficient of friction of the sliding surface is supposed to be 0.2 over the service life of the isolator. The neoprene pad has a very low displacement capacity, probably not more than ±5.0cm (about 2in.). When the displacements exceed this, the sliding element provides the needed movement. The system does not include any restoring device and permanent displacements could occur. The system has been implemented only once in a large nuclear power plant at Koeberg, South Africa.

2. EERC combined system

A combination elastomeric and sliding system was developed and tested on the shake table at the EERC. In this system, the interior columns of the structure were carried on Teflon on stainless steel sliding elements and the exterior columns on the low-damping natural rubber bearings. The elastomeric bearings provided recentering capability and controlled the torsion of the structure while the sliding elements provided damping.

A variant of this system was used to retrofit both the Mackay School of Mines at the University of Nevada, Reno, Nevada, and a new hospital for the County of Los Angeles, the M. L. King, Jr. -C. R. Drew Diagnostics Trauma Center in Willowbrook, California. Both of these structures used HDNR elastomeric bearings; teflon-stainless steel sliding elements were used in the university building while lead-bronze alloy plates on stainless steel were used for the hospital.

3. The TASS system

The TASS system was developed by the TAISEI Corp. in Japan. In this system, the entire vertical load is carried on Teflon-stainless steel elements. In addition, laminated neoprene bearings that carry no load are used to provide recentering forces. The Teflon sliding surface has a pressure of around 10MPa (1450psi), and the coefficient friction ranges from 0.05 at slow sliding speeds to around 0.15 at higher speeds. The disadvantages of this system are that because the elastomeric bearings carry no vertical load, they experience tension, and the velocity sensitivity of the sliding surface makes modeling of the system quite difficult.

4. Resilient-friction base isolation system

The resilient-friction base isolation (R-FBI) bearing attempts to overcome the problem of the high friction coefficient of Teflon on stainless steel at high velocities by using many sliding interfaces in a single bearing. Thus the velocity between the top and bottom of the bearing is divided by the number of layers so that the velocity at each face is small, maintaining a low friction coefficient (Fig. 12-9). In addition to the sliding elements, there is a central core of rubber that carries no vertical load but provides a restoring force. Tests of this system found that the rubber core did not prevent the displacement from being concentrated at a single interface; therefore, a central steel rod was inserted in the rubber core that improved the distribution of displacement among the sliding layers. A shake table experimental test program using R-FBI bearings and a five-story, 40-ton steel frame model was carried out at the EERC in 1988.

5. Friction pendulum system

The friction pendulum system (FPS) is a frictional isolation system that combines a sliding action and a restoring force by geometry. The FPS isolator, shown schematically in Fig. 12-10, has an articulated slider that moves on a stainless steel spherical surface. The side of the articulated slider in contact with the spherical surface is coated with a low-friction composite material. The other side of the slider is also spherical, coated with stainless steel, and sits in a spherical cavity, also coated with the low-friction composite material. As the slider moves over the spherical surface, it causes the supported mass to rise and provides the restoring force for the system. Friction between the articulated slider and the spherical surface generates damping in the isolators. The effective stiffness of the isolator and the isolation period of the structure is controlled by the radius of curvature of the concave surface.

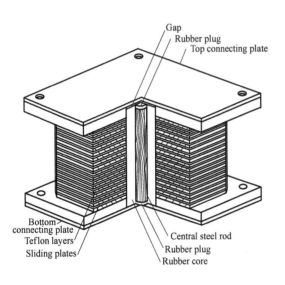

Fig. 12-9 Resilient-friction base isolation system.

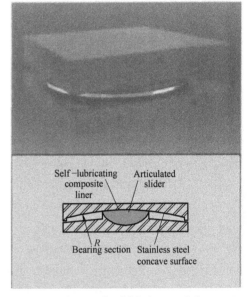

Fig. 12-10 Schematic of friction pendulum system.

12.3.3 Isolation Systems Based on Sliding Spring-type Systems

Elastomeric and sliding isolation systems are usually configured to provide only horizontal isolation. When full three-dimensional isolation is required, it is possible, but not common, to use elastomeric bearings. Generally, spring-type systems have been used in these cases.

The GERB system for seismic isolation was developed originally for the vibration isolation of power plant turbine generating equipment. It uses large helical steel springs that are flexible both horizontally and vertically.[9] The vertical frequency is around 3 ~ 5 times the horizontal frequency. The steel springs are completely without damping and the system is always used in conjunction with the GERB visco damper. As in all three-dimensional systems, there is very strong coupling between horizontal motion and rocking motion because the center of gravity of the isolated structure is above the center of stiffness of the isolation system. This type of system becomes practical in situations where the center of gravity and the center of stiffness are at the same level—in a reactor vessel in a nuclear power plant, for example.

The system has been tested on the shake table at Skopje, Macedonia, and has been implemented in two steel frame houses in Santa Monica, California (see Fig. 12-11). These houses were strongly affected by the 1994 Northridge earthquake. Their response was monitored by strong motion instruments and demonstrates that the isolation system was not effective in reducing the accelerations in these buildings due to the rocking motion.

12.3.4 Rocking Systems

Tall slender structures of top-heavy construction will inevitably develop overturning moments that will produce tension at the foundation level. It is extremely expensive to provide tension capacity in building foundations using anchors in deep caissons. As an alternative, it is possible to allow the columns—or piers in the case of bridges—to step off the foundation. This form of partial isolation reduces the seismic loads throughout the structure, particularly the tension forces generated in columns or piers.[10] The dynamics of the stepping structure are quite different from conventional structures but have been extensively studied both theoretically and experimentally in shake table tests at the EERC.

Fig. 12-11 GERB system: lowe residences.

This concept has been implemented in a railway bridge in New Zealand. The South Rangitikei River Bridge (Fig. 12-12) has 69 m (about 226 ft) long piers that are designed to lift off of the foundation under seismic loading. Inside piers are two large energy-dissipating devices that are based on the elastic-

Fig. 12-12 South Rangitikei River Bridge.

plastic torsion of mild-steel bars. These provide controlling forces to the piers, both when the pier is moving up and when it is moving down. These devices were designed and tested at the Department of Scientific and Industrial Research Physics and Engineering Laboratory in New Zealand in 1971—1972 and are the first use of energy-dissipating devices in seismic-restraint design. The method has not been used again for bridges but could be a very promising retrofit strategy for large bridges such as the Golden Gate Bridge, the Oakland Bay Bridge, and the Williamsburg Bridge. This approach was used later to isolate a tall chimney structure in Christchurch, New Zealand; in this case tapered-plate steel dampers were included.

12.4 Merits and Demerits Analysis

Seismic isolation may be looked upon as an approach towards damage prevention rather than its cure. The principal merits and demerits of an isolator system can be summarized as follows:

12.4.1 Merits

1) Because of its low stiffness, use of an isolator system leads to an increase in the natural period of a structure as compared to that for a fixed base structure. This moves the system away from the period at which ground motion contains substantial energy. This results in lower inertia forces that the structure has to withstand leading to cost saving.

2) In the first mode of vibration (which is often the predominant mode) primary displacements occur only at isolator level while the superstructure behaves almost like a rigid body.

3) The floor accelerations are reduced which result in reduced inter-storey drifts.

4) Hinge regions in fixed base structures need to have the capacity to deform into an inelastic range over many reversible cycles while maintaining adequate strength and stiffness to ensure stability and integrity of the structure. Such plastic deformations could be large resulting in significant damage to structural and non-structural components. This is minimized in an isolator protected structure.

5) In the case of a larger than assessed seismic event, the damage gets concentrated in the isolation system which can be restored relatively easily. As a result, the structure can often be commissioned into service in a short time. This is of immense importance for buildings such as hospitals and those that house emergency service providers.

6) It is an ideal and, sometimes, the only means of retrofitting buildings of historical importance.

12.4.2 Demerits

Following are some of the demerits of supporting a building on isolators:

1) An isolator system does not significantly help an already flexible structure. As a result, when used singly it has limited application for high-rise structures. However, this limitation is becoming secondary as minimizing damage to very expensive equipment (e.g. in hospital buildings

and other structures) to ensure business continuity is gaining in importance.

2) With a base isolation system, peak base displacements are large calling for adequate rattle space to be provided to prevent the structure colliding against adjacent elements.

3) Caution is called for in the design of buildings with high aspect ratio as there could be net uplift forces on isolators brought about by large overturning moments.

4) It is less effective for a building supported on soft soil.

5) Since isolation systems are vertically stiff, vertical acceleration amplification is not prevented. Thus, it is advisable to protect equipment through secondary systems to guard against vertical earthquake motion.

6) Special flexible joints have to be built into supply lines to sustain the displacements while transiting across the isolator interface. Also rigid structures crossing the interface (e.g. stairs, walls) should have the capability to absorb lateral displacements.

7) There is additional expenditure towards foundation costs.

8) The P-β moment on an isolator can be large because of isolator displacement.

This moment will need to be considered at both top and bottom of the isolator interface.

New Words and Expressions

    seismic-resistant design　抗震设计
    inelastic　*adj.*　无弹性的，非弹性的
    deformation　*n.*　变形
    inertia　*n.*　惯性
    earthquake　*n.*　地震
    strong-column weak-beam　强柱弱梁
    layout　*n.*　布局；设计；安排；陈列
    vulnerability　*n.*　易损性
    isolation　*n.*　隔震垫，隔震器
    rigid body　刚体
    ductile　*adj.*　易延展的
    precaution　*n.*　预防，警惕；预防措施
    elevator shaft　升降机井，电梯井
    framework　*n.*　框架
    flexible　*adj.*　灵活的；柔韧的；易弯曲的
    rubber isolators　橡胶隔震器
    plug　*n.*　插头；塞子；栓
    damping　*n.*　阻尼；衰减
    Teflon　*n.*　聚四氟乙烯
    pendulum　*n.*　钟摆；摆锤

Notes

1. The system decouples the building or structure from the horizontal components of the ground motion by interposing structural elements with low horizontal stiffness between the structure and the foundation.

本句难点解析：这句话的主语为 system，谓语为 decouples，with 引导的内容作为定语修饰 elements。

本句大意如下：基础隔震系统通过在结构和基础间插入低水平刚度的结构构件，减小了地面运动水平分量对结构的影响。

2. Isolators do substantially enhance the fundamental period of a stiff structure but at the expense of increased building displacements.

本句难点解析：这句话的主语为 isolators，do 为助动词，谓语为 enhance，but 后为定语修饰 isolators。

本句大意如下：设置隔震器可以增加刚性结构的基本周期，但与此同时也会使建筑的位移增大。

3. The bearings have a vertical stiffness that is only a few times the horizontal stiffness and the rubber is relatively undamped.

本句难点解析：这句话的主语为 bearings 和 rubber，谓语分别为 have 和 is，that 引导定语从句修饰 stiffness。

本句大意如下：支座的竖向刚度仅为水平刚度的几倍，同时橡胶是相对无阻尼的。

4. The advantages of the low-damping elastomeric laminated bearings are many: They are simple to manufacture (the compounding and bonding process to steel is well understood), easy to model, and their mechanical response is unaffected by rate, temperature, history, or aging.

本句难点解析：这句话的主语为 advantages。

本句大意如下：低阻尼弹性层压支座有许多优点，它们易于制造（组分及其与钢板的粘结过程已被充分理解），易于模拟，力学响应不受速率、温度、时间和老化的影响。

5. Lead-plug bearings are laminated rubber bearings similar to low-damping rubber bearings but contain one or more lead plugs that are inserted into holes.

本句难点解析：这句话的主语是 lead-plug bearings，谓语是 are，similar 后内容修饰 bearings，that 引导定语从句修饰 plugs。

本句大意如下：铅芯橡胶支座是和低阻尼橡胶支座类似的层压橡胶支座，但它包含一个或多个嵌入孔洞的铅芯。

6. The damping is increased by adding extrafine carbon block, oils or resins, and other proprietary fillers.

本句难点解析：这句话的主语是 damping，谓语是 is increased。

本句大意如下：添加精密的碳块、油或树脂以及其他专用填料可以增大阻尼。

7. In fact, Calantarients's system incorporated all the elements now considered necessary in a base isolation system: a method of decoupling the building and the foundation, a method whereby u-

tility lines can withstand large relative displacements, and a wind restraint system.

本句难点解析：这句话的主语是 system，谓语是 incorporated，now 后是定语从句修饰 elements。

本句大意如下：事实上，Calantarients 的体系包含了所有现今被认为在基础隔震中有效的元素：分离建筑与基础的方法，管线能承受较大相对位移的方法，以及抗风体系。

8. In a sliding isolator, since the frictional force generated is proportional to the weight it supports, the center of mass and center of resistance naturally coincide.

本句难点解析：这句话的主语是 the center，谓语是 coincide，since 引导的是状语从句。

本句大意如下：在滑动隔震器中，产生的摩擦力和支承的重力成比例，因此质量和抗力的中心自然是一致的。

9. It uses large helical steel springs that are flexible both horizontally and vertically.

本句难点解析：这句话的主语是 it，it 指上一句的 GERB 隔震系统，谓语是 uses，that 后引导定语从句修饰 springs。

本句大意如下：它采用了在水平和竖直方向变形均较灵活的大型螺旋钢弹簧。

10. This form of partial isolation reduces the seismic loads throughout the structure, particularly the tension forces generated in columns or piers.

本句难点解析：这句话的主语是 form，谓语是 reduces，particularly 之后为定语从句修饰 seismic loads。

本句大意如下：这种局部隔震的形式减小了整个结构的地震荷载，特别是在柱子和桥墩上的拉力。

Exercises

Translate the following phrases into Chinese.

1. inelastic deformation capacity
2. short columns
3. seismic ground motion
4. high natural fundamental frequency
5. elasto-plastic characteristics
6. isolator system
7. hinge regions in fixed base structures
8. high aspect ratio
9. absorb lateral displacements
10. isolator interface

Translate the following sentences into Chinese.

1. The conventional approach to seismic-resistant design is to incorporate adequate strength, stiffness and inelastic deformation capacity into the building structure so that it can withstand induced inertia forces.

2. It is a sophisticated practical solution to improve seismic response of a building by

minimizing the structural damage, which was earlier taken to be unavoidable during strong ground motion.

3. Base isolation is now a mature technology and is used in many countries, and there are a number of acceptable isolation systems, the construction of which is well understood.

4. In the first mode of vibration (which is often the predominant mode) primary displacements occur only at isolator level while the superstructure behaves almost like a rigid body.

5. With a base isolation system, peak base displacements are large calling for adequate rattle space to be provided to prevent the structure colliding against adjacent elements.

Unit 13

Supplemental Energy Dissipation: State-of-the-art and State-of-the-practice

▌13.1 Introduction

In recent years, innovative means of enhancing structural functionality and safety against natural and man-made hazards have been in various stages of research and development. By and large, they can be grouped into three broad areas as shown in Table 13-1.

1) Base isolation.
2) Passive energy dissipation.
3) Active control.

Of the three, base isolation can now be considered a more mature technology with wider applications as compared with the other two.

Passive energy dissipation systems encompass a range of materials and devices for enhancing damping, stiffness and strength, and can be used both for seismic hazard mitigation and for rehabilitation of aging or deficient structures. In general, such systems are characterized by their capability to enhance energy dissipation in the structural systems in which they are installed. These devices generally operate on principles such as frictional sliding, yielding of metals, phase transformation in metals, deformation of viscoelastic (VE) solids or fluids and fluid orificing.[1]

Active, hybrid and semi-active structural control systems are a natural evolution of passive control technologies. The possible use of active control systems and some combinations of passive and active systems as a means of structural protection against seismic loads has received considerable attention in recent years.

Active/hybrid/semi-active control systems are force delivery devices integrated with real-time processing evaluators/controllers and sensors within the structure.[2] They act simultaneously with the hazardous excitation to provide enhanced structural behavior for improved service and safety. Research to date has also reached the stage where active systems have been installed in full-scale structures for seismic hazard mitigation.

This paper provides an assessment of the state-of-the-art and state-of-the-prac-

主动控制简介

tice of this exciting, and still evolving, technology. Also, included in the discussion are some basic concepts, the types of structural control systems being used and deployed, and their advantages and limitations in the context of seismic design and retrofit of civil engineering structures.

Table 13-1　Structural protective systems

Seismic isolation	PED	Semi-active and active control
Elastomeric bearings	Metallic dampers	Active bracing systems
	Friction dampers	Active mass dampers
Lead rubber bearings	VE dampers	Variable stiffness or damping systems
	Viscous fluid dampers	Smart materials
Sliding friction pendulum	Tuned mass dampers	
	Tuned liquid dampers	

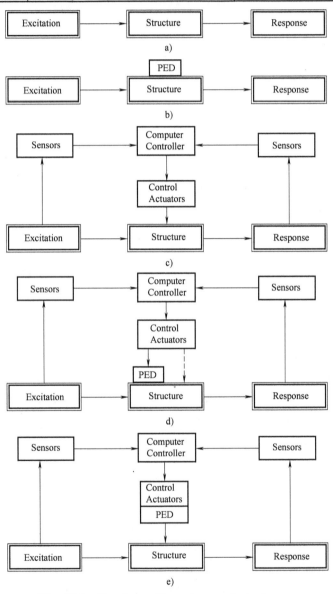

Fig. 13-1　Structure with various control schemes.

a) Conventional Structure　b) Structure with Passive Energy Dissipation (PED)　c) Structure with Active Control
d) Structure with Hybrid Control　e) Structure with Semi-Active Control

13.2 Basic Principles

In what follows, basic principles of passive and active control are illustrated using a simple single-degree-of-freedom (SDOF) structural model. Consider the lateral motion of the SDOF model consisting of a mass m, supported by springs with total linear elastic stiffness k, and a damper with damping coefficient c. This SDOF system is then subjected to an earthquake load where $\ddot{x}_g(t)$ is ground acceleration. The excited model responds with a lateral displacement $x(t)$ relative to the ground which satisfies the equation of motion (schematically represented by Fig. 13-1a)

$$m\ddot{x} + c\dot{x} + kx = -m\ddot{x}_g \tag{13-1}$$

Consider now the addition of a generic passive energy dissipation (PED) element into the SDOF model. The equation of motion for the extended SDOF model then becomes (schematically represented by Fig. 13-1b)

$$m\ddot{x} + c\dot{x} + kx + \Gamma x = -(m+\overline{m})\ddot{x}_g \tag{13-2}$$

where $\overline{m}$ is the mass of the PED element and the force corresponding to the device is written as Γx, Γ representing a generic integrodifferential operator.

The specific form of Γx needs to be specified before Eq. (13-2) can be analyzed, which is necessarily highly dependent on the device type. It is seen from Eq. (13-2) that the addition of the Γx term in Eq. (13-2) modified the structural properties so that it can respond more favorably to the designed or anticipated ground motion. It is important to note that a passive structure with added PED elements is again a passive structure.

An active structural control system, on the other hand, has the basic configuration as shown schematically in Fig. 13-1c. It consists of:

1) Sensors located about the structure to measure either external excitations, or structural response variables, or both.

2) Devices to process the measured information and to compute necessary control forces needed based on a given control algorithm.

3) Actuators, usually powered by external sources, to produce the required forces.

When only the structural response variables are measured, the control configuration is referred to as feedback control since the structural response is continually monitored and this information is used to make continual corrections to the applied control forces. A feedforward control results when the control forces are regulated only by the measured excitation, which can be achieved, for earthquake inputs, by measuring accelerations at the structural base.[3] In the case where the information on both the response quantities and excitation are utilized for control design, the term feedback-feedforward control is used.

To see the effect of applying such control forces to the linear structure considered above, Eq. (13-1) in this case becomes

$$m\ddot{x} + c\dot{x} + kx = -mu(t) - m\ddot{x}_g \tag{13-3}$$

where $u(t)$ is the applied control force.

Suppose that the feedback configuration is used in which the control force $u(t)$ is designed to be

$$u(t) = \frac{\Gamma x}{m} \tag{13-4}$$

and Eq. (13-3) becomes

$$m\ddot{x} + c\dot{x} + kx + \Gamma x = -m\ddot{x}_g \tag{13-5}$$

It is seen that the effect of feedback control is again to modify the structural properties. In comparison with passive control, however, an important difference is that the form of Γx is now governed by the control law chosen for a given application, which can change as a function of the excitation. Other advantages associated with active control systems can be cited; among them are:

1) Enhanced effectiveness in response control; the degree of effectiveness is, by and large, only limited by the capacity of the control systems.

2) Relative insensitivity to site conditions and ground motion.

3) Applicability to multi-hazard mitigation situations; an active system can be used, for example, for motion control against both strong wind and earthquakes.

4) Selectivity of control objectives; one may emphasize, for example, human comfort over other aspects of structural motion during noncritical times, whereas increased structural safety may be the objective during severe dynamic loading.

While this description of active control is conceptually in the domain of familiar optimal control theory used in electrical engineering, mechanical engineering, and aerospace engineering, structural control for civil engineering applications has a number of distinctive features, largely due to implementation issues, that set it apart from the general field of feedback control. In particular, when addressing civil engineering structures, there is considerable uncertainty, including nonlinearity, associated with both physical properties and disturbances such as earthquakes and wind, the scale of the forces involved can be quite large, there are only a limited number of sensors and actuators, the dynamics of the actuators can be quite complex, the actuators are typically very large, and the systems must be fail-safe.

It is useful to distinguish among several types of active control systems currently being used in practice. The term *hybrid control* generally refers to a combined passive and active control system as depicted in Fig. 13-1d. Since a portion of the control objective is accomplished by the passive system, less active control effort, implying less power resource, is required.

Similar control resource savings can be achieved using the semi-active control scheme sketched in Fig. 13-1e, where the control actuators do not add mechanical energy directly to the structure, hence bounded-input/bounded-output stability is guaranteed. Semi-active control devices are often viewed as controllable passive devices.

A side benefit of hybrid and semi-active control systems is that, in the case of a power failure, the passive components of the control still offer some degree of protection, unlike a fully active control system.

13.3 Passive Energy Dissipation

A large number of passive control systems or PED devices have been developed and installed in structures for performance enhancement under earthquake loads. In North America, PED devices have been implemented in approximately 103 buildings and many bridges, either for retrofit or for new construction. Discussions presented below are centered around some of the more common devices which have found applications in PED.

13.3.1 Metallic Yield Dampers

One of the effective mechanisms available for the dissipation of energy input to a structure from an earthquake is through inelastic deformation of metals. Many of these devices use mild steel plates with triangular or X shapes so that yielding is spread almost uniformly throughout the material. A typical X-shaped plate damper or ADAS (added damping and stiffness) device is shown in Fig. 13-2. Other configurations of steel yielding devices, used mostly in Japan, include bending type of honeycomb and slit dampers and shear panel type. Other materials, such as lead and shape-memory alloys, have also been evaluated. Some particularly desirable features of these devices are their stable hysteretic behavior, low-cycle fatigue property, long term reliability, and relative insensitivity to environmental temperature. Hence, numerous analytical and experimental investigations have been conducted to determine these characteristics of individual devices.

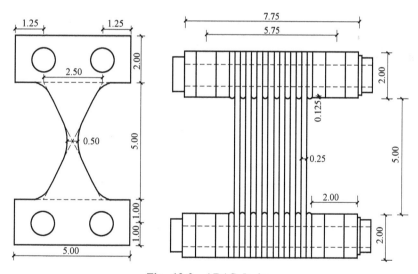

Fig. 13-2 ADAS device.
Note: All dimensions in inches.

After gaining confidence in their performance based primarily on experimental evidence, implementation of metallic devices in full-scale structures has taken place. The earliest implementations of metallic dampers in structural systems occurred in New Zealand and Japan. A number of these interesting applications are reported. More recent applications include the use of ADAS dampers in the seismic upgrade of existing buildings in Mexico and in the USA. The seismic upgrade project discussed in the USA involves the retrofit of a Wells Fargo Bank building in San Francisco, CA. The building is a two-story nonductile concrete frame structure originally constructed in 1967 and subsequently damaged in the 1989 Loma Prieta earthquake. A total of seven ADAS devices were employed, each with a yield force of 150kips. Both linear and nonlinear analyses were used in the retrofit design process. Further, three-dimensional response spectrum analyses, using an approximate equivalent linear representation for the ADAS elements, furnished a basis for the redesign effort. The final design was verified with DRAIN-2D nonlinear time history analyses. A comparison of computed response before and after the upgrade is shown in Fig. 13-3. The numerical results indicated that the revised design was stable and that all criteria were met. In addition to the introduction of the bracing and ADAS dampers, several interior columns and a shear wall were strengthened.

A variation of the devices described above but operating on the same metallic yielding principle is the tension/compression yielding brace, also called the unbonded brace, which has found applications in Japan and the USA.[4] As shown in Fig. 13-4, an unbonded brace is a bracing member consisting of a core steel plate encased in a concrete-filled steel tube. A special coating is provided between the core plate and concrete in order to reduce friction. The core steel plate provides stable energy dissipation by yielding under reversed axial loading, while the surrounding concrete-filled steel tube resists compression buckling.

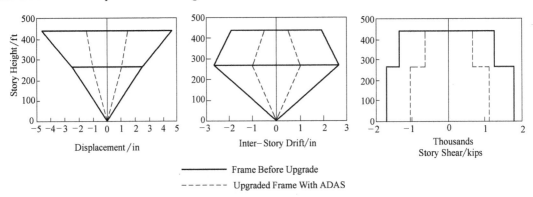

Fig. 13-3 Comparison of computed results for Wells Fargo Bank
Building—envelope of response values in the X-direction.

13.3.2 Friction Dampers

Friction dampers utilize the mechanism of solid friction that develops between two solid bodies sliding relative to one another to provide the desired energy dissipation. Several types of friction dampers have been developed for the purpose of improving seismic response of structures. An example of such a device is depicted in Fig. 13-5. During cyclic loading, the mechanism enforces slippage in both tensile and compressive directions. Generally, friction devices generate rectangular hysteretic loops similar to the characteristics of Coulomb friction. After a hysteretic restoring force model has been validated for a particular device, it can be readily incorporated into an overall structural analysis.[5]

In recent years, there have been a number of structural applications of friction dampers aimed at providing enhanced seismic protection of new and retrofitted structures. This activity in North America is primarily associated with the use of Pall friction devices in Canada and the USA; and slotted-bolted connection in the USA. For example, the applications of friction dampers to the McConnel Library of the Concordia University in Montreal, Canada are discussed. A total of 143 dampers were employed in this case. A series of nonlinear DRAIN-TABS analyses were utilized to establish

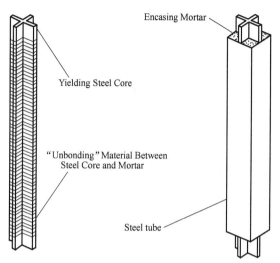

Fig. 13-4 Unbonded brace.

the optimum slip load for the devices, which ranges from 600~700 kN depending upon the location within the structure. For the three-dimensional time-history analyses, artificial seismic signals were generated with a wide range of frequency contents and a peak ground acceleration scaled to 0.18g to represent expected ground motion in Montreal. Under this level of excitation, an estimate of the equivalent damping ratio for the structure with frictional devices is about 50%. In addition, for this library complex, the use of the friction dampers resulted in a net savings of 1.5% of the total building cost.

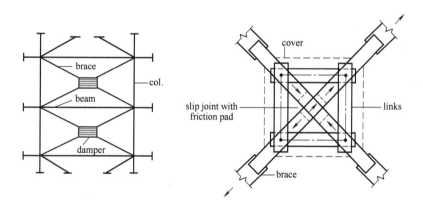

Fig. 13-5 X-braced friction damper.

13.3.3 Viscoelastic Dampers

Viscoelastic (VE) materials used in structural applications are usually copolymers or glassy substances that dissipate energy through shear deformation. A typical VE damper, which consists of VE layers bonded with steel plates, is shown in Fig. 13-6. When mounted in a structure, shear deformation and hence energy dissipation takes place when structural vibration induces relative motion between the outer steel flanges and the center plates.[6] Significant advances in research and development of VE dampers, particularly for seismic applications, have been made in recent years through analyses and experimental tests.

A seismic retrofit project using VE dampers began in 1993 for the 13-story Santa Clara County building in San Jose, CA. Situated in a high seismic risk region, the building was built in 1976. It is ca 64m in height and nearly square in plan, with 51m× 51m on typical upper floors. The exterior cladding consists of full-height glazing on two sides and metal siding on the other two sides. The exterior cladding, however, provides little resistance to structural drift. The equivalent viscous damping in the fundamental mode was <1% of critical.

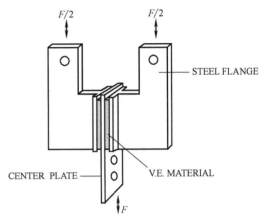

Fig. 13-6 Typical VE damper configuration.

The building was extensively instrumented, providing invaluable response data obtained during a number of past earthquakes. A plan for seismic upgrade of the building was developed, in part, when the response data indicated large and long-duration response, including torsional coupling, to even moderate earthquakes. The final design called for installation of two dampers per building face per floor level, which would increase the equivalent damping in the fundamental mode of the building to about 17% of critical, providing substantial reductions to building response under expected levels of ground shaking. A typical damper configuration is shown in Fig. 13-7. More recent installations include the use of VE dampers to upgrade a concrete structure and their use in a new construction.

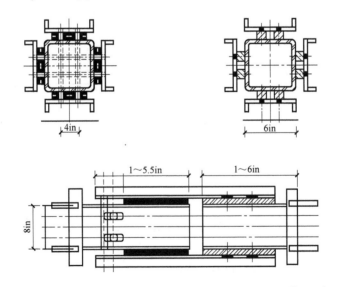

Fig. 13-7 Santa Clara County Building-VE damper configuration
(longitudinal and cross-sectional views).

In Japan, the Hazama Corp. developed similar devices by using similar materials, and the Shimizu Corp developed VE walls, in which solid thermoplastic rubber sheets were sandwiched between steel plates.

13.3.4 Viscous fluid dampers

The viscous fluid (VF) devices developed recently include viscous walls and VF dampers. The viscous wall, developed by Sumitomo Construction Company, consists of a plate moving in a thin steel case filled with highly VF. The VF damper, widely used in the military and aerospace industry for many years, has recently been adapted for structural applications in civil engineering. <u>A VF damper generally consists of a piston within a damper housing filled with a compound of silicone or similar type of oil, and the piston may contain a number of small orifices through which the fluid may pass from one side of the piston to the other.</u>[7] Thus, VF dampers dissipate energy through the movement of a piston in a highly VF based on the concept of fluid orificing.

VF dampers have in recent years been incorporated into a large number of civil engineering structures. In several applications, they were used in combination with seismic isolation systems. For example, in 1995, VF dampers were incorporated into base isolation systems for five buildings of the San Bernardino

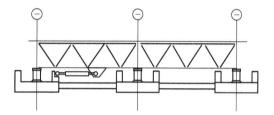

Fig. 13-8 San Bernardino County Medical Center-damper-base isolation system assembly.

County Medical Center, located close to two major fault lines. The five buildings required a total of 233 dampers, each having an output force capacity of 320000lb and generating an energy dissipation level of 3000hp (1hp=745.700W) at a speed of 60 in/s. A layout of the damper-isolation system assembly is shown in Fig. 13-8 and Fig. 13-9 gives the dimensions of the viscous dampers employed.

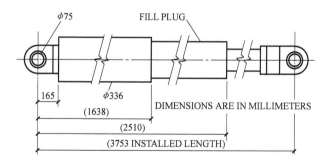

Fig. 13-9 Dimensions of VF damper for San Bernardino County Medical Center.

13.3.5 Tuned Mass Dampers

Early applications of tuned mass dampers (TMDs) have been directed toward

TMD

mitigation of wind-induced excitations. Recently, numerical and experimental studies have been carried out to examine the effectiveness of TMDs in reducing seismic response of structures. It is noted that a passive TMDs can only be tuned to a single structural frequency. While the first-mode response of a MDOF structure with TMDs can be substantially reduced, the higher mode response may in fact increase as the number of stories increases. For earthquake-type excitations, the response reduction is large for resonant ground motions and diminishes as the dominant frequency of the ground motion gets further away from the structure's natural frequency to which the TMDs is tuned.

It is also noted that the interest in using TMDs for vibration control of structures under earthquake loads has resulted in some innovative developments. An interesting approach is the use of TMDs with active capability, the so called active mass damper (AMD) or hybrid mass damper (HMD). Systems of this type have been implemented in a number of tall buildings in recent years in Japan, and they are described in the next section.

13.3.6 Tuned Liquid Dampers

The basic principles involved in applying a tuned liquid dampers (TLDs) to reduce the dynamic response of structures is quite similar to that discussed above for the TMDs. In effect, a secondary mass in the form of a body of liquid is introduced into the structural system and tuned to act as a dynamic vibration absorber. However, in the case of TLDs, the damper response is highly nonlinear due either to liquid sloshing or the presence of orifices. TLDs have also been used for suppressing wind-induced vibrations of tall structures. In comparison with TMDs, the advantages associated with TLDs include low initial cost, virtually free of maintenance and ease of frequency tuning.

The TLD applications have taken place primarily in Japan for controlling wind-induced vibration. Examples of TLD-controlled structures include airport towers and tall buildings.

13.4 Active, Hybrid and Semi-active Control Systems

The rapid growth of research interest and development of active/hybrid and semi-active structural control systems is in part due to several coordinated research efforts, largely in Japan and the USA, marked by a series of milestones listed in Table 13-2. Indeed, the most challenging aspect of active control research in civil engineering is the fact that it is an integration of a number of diverse disciplines, some of which are not within the domain of traditional civil engineering. These include computer science, data processing, control theory, material science, sensing technology, as well as stochastic processes, structural dynamics, and wind and earthquake engineering. These coordinated efforts have facilitated collaborative research efforts among researchers from diverse backgrounds and accelerated the research-to-implementation process as one sees today.

Table 13-2 Active structural control research—milestones

Year	Event
1989	US Panel on Structural Control Research (US-NSF)

(Continued)

Year	Event
1990	Japanese Panel on Structural Response Control (Japan-SCJ)
1991	Five-year Research Initiative on Structural Control (US-NSF)
1993	European Association for Control of Structures
1994	International Association for Structural Control
1994	First World Conference on Structural Control (Pasadena, CA, USA)
1996	First European Conference on Structural Control (Barcelona, Spain)
1998	Chinese Panel for Structural Control
1998	Korean Panel for Structural Control
1998	Second World Conference on Structural Control (Kyoto, Japan)
2000	Second European Conference on Structural Control (Paris, France)
2002	Third World Conference on Structural Control (Como, Italy)

As alluded to earlier, the development of active, hybrid, and semi-active control systems has reached the stage of full-scale applications to actual structures. Other than these installations in building structures and towers, most of which are in Japan, 15 bridge towers have employed active systems during erection in addition. Most of these full-scale systems have been subjected to actual wind forces and ground motions and their observed performances provide invaluable information in terms of:

1) Validating analytical and simulation procedures used to predict actual system performance.

2) Verifying complex electronic-digital-servo hydraulic systems under actual loading conditions.

3) Verifying capability of these systems to operate or shutdown under prescribed conditions.

Described below are several of these systems together, in some cases, with their observed performances. Also addressed are several practical issues in connection with actual structural applications of these systems.

13.4.1 Hybrid Mass Damper Systems

HMD 系统

The HMD is the most common control device employed in full-scale civil engineering applications. HMD is a combination of passive TMDs and an active control actuator. The ability of this device to reduce structural responses relies mainly on the natural motion of the TMDs. The forces from the control actuator are employed to increase the efficiency of the HMD and to increase its robustness to changes in the dynamic characteristics of the structure. The energy and forces required to operate a typical HMD are far less than those associated with a fully AMP system of comparable performance.

An example of such an application is the HMD system installed in the Sendagaya INTES Building in Tokyo in 1991. As shown in Fig. 13-10, the HMD was installed atop the 11th floor and consists of two masses to control transverse and torsional motions of the structure, while hydraulic actuators provide the active control capabilities. The top view of the control system is shown in Fig. 13-11 where ice thermal storage tanks are used as mass blocks so that no extra mass was introduced. The masses are supported by multi-stage rubber bearings intended for reducing the control energy consumed in the HMD and for insuring smooth mass movements.

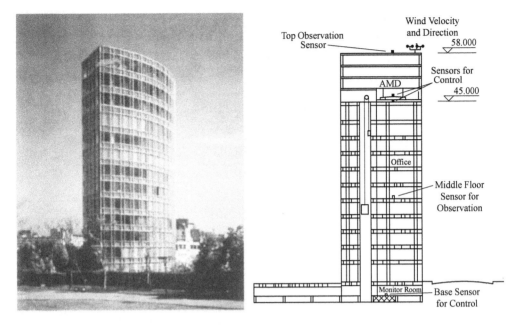

Fig. 13-10　Sendagaya INTES Building with hybrid mass dampers.

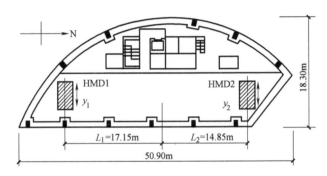

Fig. 13-11　Top view of HMD configuration.

Variations of such a HMD configuration include multi-step pendulum HMD (as seen in Fig. 13-12), which have been installed in, for example, the Yokohama Landmark Tower in Yokohama, the tallest building in Japan, and in the TC Tower in Kaohsiung, Taiwan, China. Additionally, the DUOX HMD system which, as shown schematically in Fig. 13-13, consists of a TMD actively controlled by an auxiliary mass, has been installed in, for example, the Ando Nishikicho Building in Tokyo.

13.4.2　Active Mass Damper Systems

Design constraints, such as severe space limitations, can preclude the use of a HMD system. Such is the case in the active mass damper or active mass driver (AMD) system designed and installed in the Kyobashi Seiwa Building in Tokyo and the Nanjing Communication Tower in Nanjing, China.

Unit 13 Supplemental Energy Dissipation: State-of-the-art and State-of-the-practice

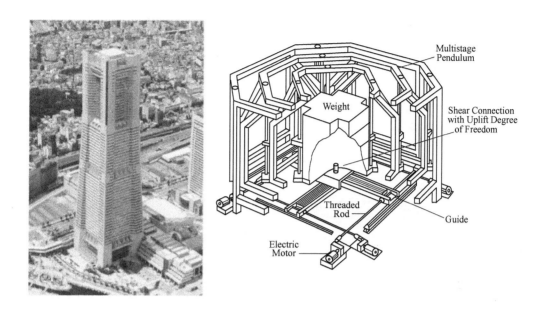

Fig. 13-12 Yokohama Landmark Tower and HMD.

The Kyobashi Seiwa Building, the first full-scale implementation of active control technology, is an 11-story building with a total floor area of 423m². As seen in Fig. 13-14, the control system consists of two AMDs where the primary AMD is used for transverse motion and has a weight of 4 ton, while the secondary AMD has a weight of 1 ton and is employed to reduce torsional motion. The role of the active system is to reduce building vibration under strong winds and moderate earthquake excitations and consequently to increase comfort of occupants in the building.

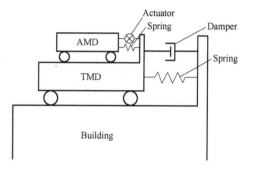

Fig. 13-13 Principle of DUOX system.

In the case of the Nanjing Communication Tower (Fig. 13-15), numerous physical constraints had to be accounted for in the system design of the mass damper. The physical size of the damper was constrained to a ring-shaped floor area with inner and outer radii of 3m and 6.1m, respectively. In addition, the damper was by necessity elevated off the floor on steel supports with Teflon bearings to allow free access to the floor area.[8] The final ring design allowed the damper to move ±750mm from its rest position. Simulations indicate that this stroke is sufficient to control the tower; however, a greater stroke would allow substantially more improvement in the response. The strength of the observation deck limited the weight of the damper to 60 ton. Lack of sufficient lateral space made the use of mechanical springs impractical for restoring forces. Thus the active control actuators

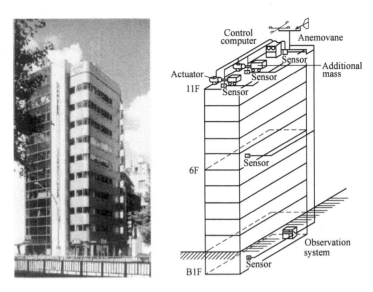

Fig. 13-14　Kyobashi Seiwa Building and AMD.

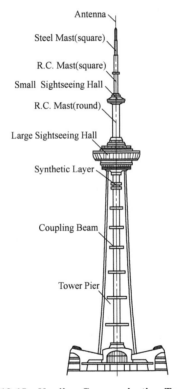

Fig. 13-15　Nanjing Communication Tower.

provide restoring force as well as the damping control forces.

　　The final design of the AMD is shown in Fig. 13-16, which uses three servo-controlled hydraulic actuators, each with a total stroke of ±1.50m and a peak control force of 50kN. These actuators are arranged 120° apart around the circumference of the ring. The actuators control three degrees of freedom: two orthogonal lateral directions of motion and torsional rotation, which is held to zero. Since the frictional force between the Teflon bearings and the mass can have a critical influence on the response of the system, a detailed analysis was performed to verify system performance in the presence of friction.

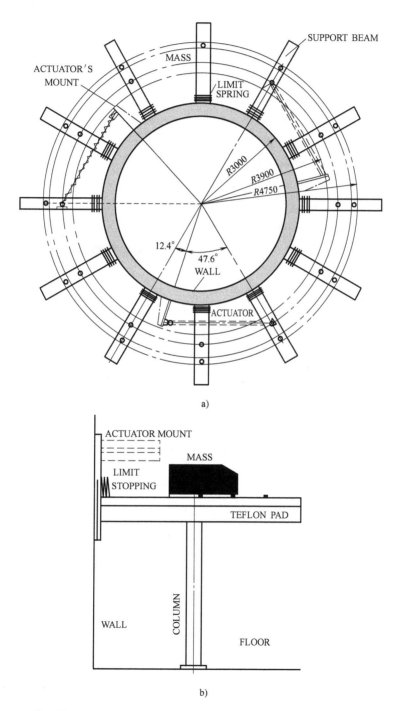

Fig. 13-16 Design of AMD showing the mass ring and actuators.

13.4.3 Semi-active Damper Systems

Control strategies based on semi-active devices combine the best features of both passive and active control systems. The close attention received in this area in recent years can be attributed to the fact that semi-active control devices offer the adaptability of active control devices without requiring the associated large power sources. In fact, many can operate on battery power, which is critical during seismic events when the main power source to the structure may fail. In addition, as stated earlier, semi-active control devices do not have the potential to destabilize (in the bounded input/bounded output sense) the structural system. Extensive studies have indicated that appropriately implemented semi-active systems perform significantly better than passive devices and have the potential to achieve the majority of the performance of fully active systems, thus allowing for the possibility of effective response reduction during a wide array of dynamic loading conditions.

One means of achieving a semi-active damping device is to use a controllable, electromechanical, variable-orifice valve to alter the resistance to flow of a conventional hydraulic fluid damper. A schematic of such a device is given in Fig. 13-17. Experiments were conducted in which a hydraulic actuator with a controllable orifice was implemented in a single-lane model bridge to dissipate the energy induced by vehicle traffic (Fig. 13-18), followed by a full-scale experiment conducted on a bridge on interstate highway I-35 to demonstrate this technology, as shown in Fig. 13-19.[9] This experiment constitutes the first full-scale implementation of active structural control in the USA.

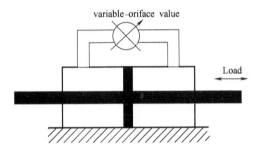

Fig. 13-17 Schematic of variable-orifice damper.

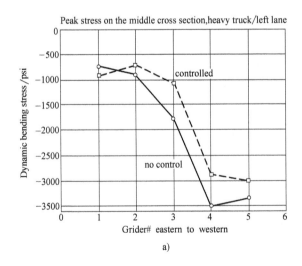

Fig. 13-18 Comparison of peak stresses for heavy trucks.

Unit 13 Supplemental Energy Dissipation: State-of-the-art and State-of-the-practice

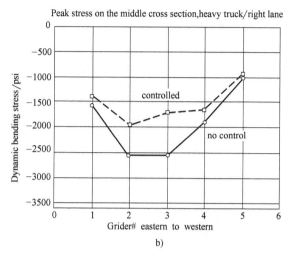

Fig. 13-18 Comparison of peak stresses for heavy trucks. (Continued)

More recently, a semi-active damper system was installed in the Kajima Shizuoka Building in Shizuoka, Japan. As seen in Fig. 13-20 semi-active hydraulic dampers are installed inside the walls on both sides of the building to enable it to be used as a disaster relief base in post-earthquake situations. Each damper contains a flow control valve, a check valve and an accumulator, and can develop a maximum damping force of 1000kN. Fig. 13-21 shows a sample of the response analysis results based on one of the selected control schemes and several earthquake input motions with a scaled maximum velocity of 50cm/s, together with a simulated Tokai wave. It is seen that both story shear forces and story drifts are greatly reduced with control activated. In the case of the shear forces, they are confined within their elastic-limit values (indicated by the E-limit in Fig. 13-21) while, without control, they would enter the plastic range.

Fig. 13-19 Highway I-35 Bridge with semi-active dampers.

13.4.4 Semi-active Controllable Fluid Dampers

Another class of semi-active devices uses controllable fluids, schematically shown in Fig. 13-22. In comparison with semi-active damper systems described above, an advantage of controllable fluid devices is that they contain no moving parts other than the piston, which makes them simple and potentially very reliable.

Two fluids that are viable contenders for development of controllable dampers are:
1) Electrorheological (ER) fluids.
2) Magnetorheological (MR) fluids.

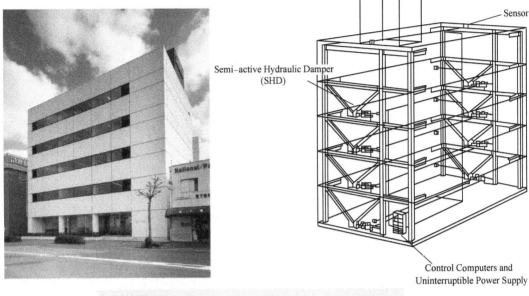

Fig. 13-20 Kajima Shizuoka Building and semi-active hydraulic dampers.

The essential characteristic of these fluids is their ability to reversibly change from a free-flowing, linear VF to a semi-solid with a controllable yield strength in milliseconds when exposed to an electric (for ER fluids) or magnetic (for MR fluids) field. In the absence of an applied field, these fluids flow freely and can be modeled as Newtonian. When the field is applied, a Bingham plastic model is often used to describe the fluid behavior. In this model, the plastic viscosity is defined as the slope of the measured shear stress versus shear strain rate data. Thus, the total yield stress is given by

$$\tau = \tau_{y(\text{field})} \operatorname{sgn}(\dot{\gamma}) + \eta_p \dot{\gamma} \qquad (13\text{-}6)$$

where $\tau_{y(\text{field})}$ is the yield stress caused by the applied field, $\dot{\gamma}$ is the shear strain rate and η_p is the plastic viscosity, defined as the slope of the measured shear stress versus shear strain rate data.

Although the discovery of both ER and MR fluids dates back to the late 1940s, for many years research programs concentrated primarily on ER fluids. Nevertheless, some obstacles remain in the development of commercially feasible damping devices using ER fluids. For example, the best ER fluids currently available have a yield stress of only 3.0~3.5kPa and cannot tolerate common impurities (e.g., water) that might be introduced during manufacturing or use. In addition, safety, availability and cost of the high voltage (e.g., ~4000V) power supplies required to control the ER fluids need to be addressed.

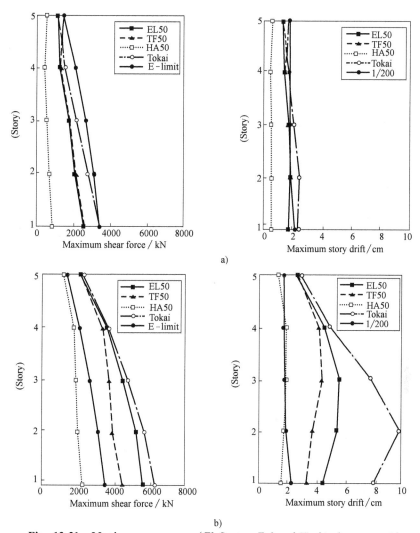

Fig. 13-21 **Maximum responses** (El Centro, Taft and Hachinohe waves with 50cm/s and assumed Tokai waves).

a) With SAHD control b) Without control

Recently developed MR fluids appear to be an attractive alternative to ER fluids for use in controllable fluid dampers. MR fluids typically consist of micronsized, magnetically polarizable particles dispersed in a carrier medium such as mineral or silicone oil.[10] It is

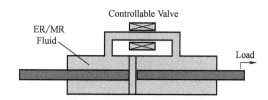

Fig. 13-22 **Schematic of controllable fluid damper**

indicated that the achievable yield stress of a MR fluid is an order of magnitude greater than its ER counterpart and that MR fluids can operate at temperatures from −40℃ to 150℃ with only modest variations in the yield stress. Moreover, MR fluids are not sensitive to impurities such as those commonly encountered during manufacturing and usage, and little particle/carrier fluid separation takes place in MR fluids under common flow conditions. The size, shape and performance of a given device is determined by a combination of $\tau_{y(\text{field})}$ and η_p. The design equations for most controllable damper geometries indicate that minimizing the ratio $\eta_p/\tau_{y(\text{field})}^2$ is desirable. This ratio for MR fluids ($\approx 5 \times 10^{-11}$ s/Pa) is three orders of magnitude smaller than the corresponding ratio for today's best ER fluids. Thus, controllable devices using MR fluids have the potential of being much smaller than ER devices with similar capabilities. Further, the MR fluid can be readily controlled with a low power (e.g., <50W), low voltage (e.g., ~12~24V), current-driven power supply outputting only ~1~2 amps. Batteries can readily supply such power levels.

13.5 Concluding Remarks

An attempt has been made in this paper to introduce the basic concepts of passive and active structural control and to bring up-to-date their current development and structural applications in this exciting and fast expanding field. While significant strides have been made in terms of implementation of these concepts to structural design and retrofit, it should be emphasized that this entire technology is still evolving. Significant improvements in both hardware, software and design procedures will certainly continue for a number of years to come.

The acceptance of innovative systems in structural engineering is based on a combination of performance enhancement versus construction costs and long-term effects. Continuing efforts are needed in order to facilitate wider and speedier implementation. These include effective system integration and further development of analytical and experimental techniques by which performances of these systems can be realistically assessed. Structural systems are complex combinations of individual structural components. New innovative devices need to be integrated into these complex systems, with realistic evaluation of their performance and impact on the structural system, as well as verification of their ability for long-term operation.

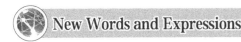

  mitigation *n.* 缓和，减少
  sensor *n.* 传感器
  retrofit *v.* 改造，翻修
  upgrade *v.* 加固
  rehabilitation *n.* 修复
  by and large *n.* 大体上，总的来说
  viscoelastic *n.* 黏弹性

Unit 13 Supplemental Energy Dissipation: State-of-the-art and State-of-the-practice

frequency content　频谱
stochastic process　随机过程
excitation　*n.*　激励
mount　*n.*　安装
slosh　*v.*　晃动
orifice　*n.*（小）孔；*v.* 从孔流出
honeycomb　*n.*　蜂窝状
deficient　*adj.*　有缺陷的
phase transformation　相变
hysteretic restoring force　滞回恢复力
metallic yield damper　金属屈服阻尼器
friction damper　摩擦阻尼器
viscoelastic damper　黏弹性阻尼器
viscous fluid damper　黏性流体阻尼器
tuned mass damper　调谐质量阻尼器
tuned liquid damper　调谐液体阻尼器
elastomeric bearings　弹性支座
lead rubber bearings　铅芯橡胶支座
sliding friction pendulum　滑移摩擦摆锤

Notes

1. These devices generally operate on principles such as frictional sliding, yielding of metals, phase transformation in metals, deformation of viscoelastic (VE) solids or fluids and fluid orificing.

本句难点解析：phase transformation in metals 指"金属相变"，在金属学中，相变常指一种组织在温度或压力变化时，转变为另一种或多种组织的过程，如多晶型转变、珠光体相变等。

本句大意如下：这些装置通常基于滑动摩擦、金属屈服、金属相变、固体液体的黏弹性变形以及孔板节流等原理来运转（消耗能量）。

2. Active/hybrid/semi-active control systems are force delivery devices integrated with real-time processing evaluators/controllers and sensors within the structure.

本句难点解析：active control systems 指"主动控制系统"，根据外界刺激和结构响应预估计所需的控制力，从而输入能量驱使作动器施加控制力或调节控制器性能参数，达到减震效果。hybrid control system：混合控制系统，在结构中联合世家主动控制和被动控制，以减少结构在外界荷载作用下的动力响应。semi-active control systems：半主动控制，以被动控制为主，只是应用少量能量对被动控制系统的工作状态进行切换，以适应系统对最优状态的跟踪。这里的 force 不是动词，而是 force delivery devices 中 devices 的定语。

本句大意如下：主动控制、混合控制和半主动控制体系是与结构中的实时处理评估/控制器共同集成的传力装置。

3. A feedforward control results when the control forces are regulated only by the measured excitation, which can be achieved, for earthquake inputs, by measuring accelerations at the structural base.

本句难点解析：which 引导非限制性定语从句，被修饰的为前面整个句子。后文的 for earthquake inputs 为插入语，不影响句子结构。

本句大意如下：当控制力仅仅通过测量的地震激励调节时，可以产生前馈控制结果。用于输入的地震激励可以通过测量结构基底的加速度获得。

4. A variation of the devices described above but operating on the same metallic yielding principle is the tension/compression yielding brace, also called the unbonded brace, which has found applications in Japan and the USA.

本句大意如下：基于同样的金属屈服原理，上述装置的一个变体是拉压屈服支撑，也称为无粘结支撑，这项技术目前在日本和美国都有应用。

5. After a hysteretic restoring force model has been validated for a particular device, it can be readily incorporated into an overall structural analysis.

本句难点解析：hysteretic restoring force model 指"恢复力特性曲线（又称滞回曲线）模型是恢复力随着变形变化的曲线"。在结构抗震分析中，恢复力模型是进行抗震分析的基础。目前多采用拟静力试验（又称往复静荷载试验）方法来确定滞回曲线。be incorporated into 指"使成为……的一部分"。

本句大意如下：在一个特定装置的滞回曲线模型已经被确定后，它可以很容易地加入到整体结构的分析当中。

6. When mounted in a structure, shear deformation and hence energy dissipation takes place when structural vibration induces relative motion between the outer steel flanges and the center plates.

本句难点解析：mount 指"安装"。这是一个由两个 when 引导的条件状语从句组成的句子。

本句大意如下：当被安装在结构中时，结构振动引起外部钢板翼缘和中心板的相对位移会导致（黏弹性材料的）剪切变形，从而产生能量的耗散。

7. A VF damper generally consists of a piston within a damper housing filled with a compound of silicone or similar type of oil, and the piston may contain a number of small orifices through which the fluid may pass from one side of the piston to the other.

本句难点解析：VF, viscous fluid (VF) devices 指"黏性液体阻尼器"。through which 后接定语从句，对 piston 活塞进行修饰。

本句大意如下：黏性液体装置通常包含一个阻尼器壳体及其内部的活塞。壳体内填充了有机硅化合物或者类似种类的油，活塞包含一些小孔，液体可以从活塞的一端流到另一端。

8. In addition, the damper was by necessity elevated off the floor on steel supports with Teflon bearings to allow free access to the floor area.

本句难点解析：by necessity 指"必然的，不可避免的"。access to 指"有……的途径进入"。

本句大意如下：此外，阻尼器必须用带有聚四氟乙烯支撑的钢结构从地面上举起，以保证（人）可以自由进入楼面。

9. Experiments were conducted in which a hydraulic actuator with a controllable orifice was implemented in a single-lane model bridge to dissipate the energy induced by vehicle traffic (Fig. 13-18),

followed by a full-scale experiment conducted on a bridge on interstate highway I-35 to demonstrate this technology, as shown in Fig. 13-19.

本句难点解析：in which 引导地点状语从句，with a controllable orifice 为定语，修饰状语从句的主语 hydraulic actuator，induced by vehicle traffic 为定语，修饰目的状语从句的宾语 energy。

本句大意如下：试验通过在单车道桥梁模型安装一个可控孔液压器来消耗车辆交通产生的能量。为验证这项技术，随后在州际公路 I-35 桥上进行足尺试验。

10. MR fluids typically consist of micronsized, magnetically polarizable particles dispersed in a carrier medium such as mineral or silicone oil.

本句难点解析：这句话的词汇比较难，句式相对简单，dispersed in 之后跟的是对 particles 进行修饰的定语从句。

本句大意如下：MR（magnetorheological 磁流变）流体通常由在载体（例如矿物油或者硅油）中散布的微米级磁性极化粒子组成。

 **Exercises**

Translate the following phrases into Chinese.

1. passive energy dissipation
2. deformation of viscoelastic (VE) solids
3. SDOF model
4. inelastic deformation of metals
5. shear panel type
6. sensing technology
7. complex electronic-digital-servo hydraulic systems
8. controllable fluid dampers
9. bring up-to-date current development
10. be realistically assessed

Translate the following sentences into Chinese.

1. Passive energy dissipation systems encompass a range of materials and devices for enhancing damping, stiffness and strength, and can be used both for seismic hazard mitigation and for rehabilitation of aging or deficient structures.

2. Consider the lateral motion of the SDOF model consisting of a mass m, supported by springs with total linear elastic stiffness k, and a damper with damping coefficient c.

3. Hence, numerous analytical and experimental investigations have been conducted to determine these characteristics of individual devices.

4. These coordinated efforts have facilitated collaborative research efforts among researchers from diverse backgrounds and accelerated the research-to-implementation process as one sees today.

5. The acceptance of innovative systems in structural engineering is based on a combination of performance enhancement versus construction costs and long-term effects.

Unit 14

Introduction to 3D Printing of Buildings and Building Components

■ 14.1 3D Printing Technology and Materials

The first 3D printer was invented in 1984 over the last decades, and 3D printing has become one of the fastest growing technologies. At the beginning, it was very complicated and what is more, expensive technology. Over the years, 3D printing started to be presented in everyday life and printers became commonly used in all kinds of industry fields. A lot of achievements have been made in medicine, automotive or aerospace industry. Thanks to the open source systems, prototyping of new product, and innovative applications of 3D printing in various fields are available for everyone. Improvement of the printing material and 3D technology became to be the goal for many companies all over the world from all industry sectors. In 2014, real revolution in construction industry has started, as the first house was printed, starting a new chapter in building technology. The questions asked in this document are: is 3D printing technology effective enough to go out of laboratory settings and be embraced by building industry? To which extent 3D printing can replace traditional construction technologies? What are the application areas? Where this technology is to be applied first?

The idea of 3D printing was born in 1983, when Charles W. Hull came up with an idea of hardening the tabletop coatings with the UV light. This simple thought has lead him to invention of stereolithography, first technology of 3D printing. Stereolithography was the first technology of rapid prototyping which means fast, precise and repeatable production of elements usually with computer support. First step in creating the technology was invention of additions to the synthetic resins that after lightening of the resins, were causing start of depolymerization process.[1] Stereolithography is a technology that can build objects with a high precision and extremely complicated geometry and that is the reason why it is used in many fields like for example: medicine, automotive and plane industry, and even art and design. Similar technique for 3D printing is selective laser sintering (SLS) in which laser is used to melt a particle of powder together to create an object. Materials used in SLS technology usually have high strength and flexibility. The most popular ones are nylon or polystyrene. Fused deposition modeling (FDM) is a technology that was invented in 1988 by S. Scott Crump. Ductile materials which are hardening itself during cooling process, are extruded through double headed nozzle.[2] Both, modelling and supportive materials are being deposited according to the cross-section layers, generated from digital model supporting the printer.[3] The nozzle contains resistive heaters that keep the filament in appropriate melting point, which allows it to flow easily through the nozzle, in case to form the layers. Like in the other technologies, after creating one layer, a platform is being lowered and next layer is created. This process is repeated until the whole object is completed. Materials usually used in FDM technology are called filaments and are used in printers as a roll of thermoplastic materials like ABS (Acrylonitrile Butadiene Styrene) or PLA (Polylactic Acid)—which is a completely different kind of thermoplastic. It's being made from corn starch or sugar cane and is biodegradable, so it is considered as greener and more sustainable than ABS. Over the past two decades, fused deposition modelling has become the most popular and widely used 3D printing method in the world. Wide range of materials were developed during the last decades presenting various properties and allowing to increase the range of applications and giving the prints the aspect of wood (PLA with wood fibers), metal (PLA with bronze), sandstone (PLA with milled chalk).

14.2 Examples of 3D Printing Building

14.2.1 Canal House in Amsterdam

In 2014, Dutch designing company Dus Architects decided to build a house by printing its parts by a giant printer. In Europe, this is the first project that will be realized entirely by 3D printing technology. Project called 3D Print Canal House takes place in Amsterdam and it is going to take at least three years. Architects from Dus Architects want to prove, that by printing components of the house directly on the site, they will be able to completely eliminate building waste and minimize costs of the transport. Mobility of the printer, is a considered as the main advantage as it may be transported all over the world, thanks to what, a cost of transport of the material and its storage on a building site will probably disappear. The time of the project was estimated, to allow them studying the technologies of the printing and developing the appropriate material. Building site is open to the public and it will remain open even after the project is finished, as the main aim of the operation is to discover and share potential usage of 3D printing in construction industry.

Components of the house are printed by a giant 3D printer called Kamer Maker. Printing technique is very similar to most of the printers. Process starts on a computer, where in a respective 3D program models are being created and converted to the desired format. Thermoplastic material (what in this particular case is biodegradable plastic), is heated by the printer until it reaches appropriate liquid state, so it can be lay down by a printer's nozzle. After one layer is created, another layer is built on the previous one. In this stage of the process the most challenging thing to develop is a material that after fabrication by the printer will be at the same time flexible enough to create fitting layers, adhesive so the subsequent layer will join with the previous one and stiff enough so that the component will preserve its shape.

14.2.2 WinSun Company Buildings

WinSun Decoration Design Engineering Co is a Chinese enterprise, working on material similar to concrete that will be suitable to use in 3D printing technology. In 2014, they have accomplished to build houses printed in 3D technology. This technology is based on building components printed as prefabricated elements and assembled on a site. Components are being printed by printer, 6 meters high, 10 meters wide and 40 meters long. The printer extrudes the material (mortar) through a nozzle layer-by-layer. Walls have diagonally reinforced pattern, with hollow structure that will be acting as insulation layer. Components are being printed in a factory and after printing, they are being transported to the building site and assembled together to create whole construction. Windows and doors were fitted in building walls (Fig. 14-1a). After roof was installed, finishing works were done and buildings were completed. The estimated cost of each building is 4800 dollars.

Year after Chinese developer has printed also five-storey building (Fig. 14-1b) using the same technique as for previous houses this building remains the tallest construction printed in 3D in the world.

a) b)

Fig. 14-1

a) First house printed by a WinSun company in 2014 b) Five-storey building printed in 3D.

14.2.3 In-situ Contour Crafting

The most promising 3D printing technology used in building industry is called Contour Crafting (CC) technology. In this technology material is poured progressively layer by layer, however whole process is taking place on site. This technique gives a great opportunity of automation of the construction process, by using 3D printer that will be able to print a whole house directly on-site. The major advantages presented by Khoshnevis are that the process that will be performed mostly by the machine, will be safer and that with use of appropriate material and with good parameters of the printer it will reduce its costs and time. 3D printing will also allow to create large components with unlimited architectural flexibility and highest precision. The idea of the inventor is to create a printer, that will have one or few nozzles that are moving on two parallel lanes installed at the construction site, separated from themselves a few meters wider than the width of the building.[4] The next part of the process is the same as in previous technologies, material is extruded through the nozzle and laid down in a shape of empty blocks, with crosswise pattern inside to ensure desired stiffness and strength.

Existing example of Contour Crafting technology realization is from Andy Rudenko's garden, where he managed to build a castle (Fig. 14-2a and b), using technology and software from RepRap 3D printing open source project. Material used in a printer was a mix of cement and sand. Whole building was printed on a single run, except of towers, that were printed separately and assembled to the building.

Fig. 14-2

a) First structure printed in-situ   b) Printing progress

14.2.4 Material Issues in 3D Printing of Building Components

Technology of Fused Deposition Modeling used in Canal House requires material development. Finding appropriate material for this technology remains the biggest challenge, in building projects involving 3D technique. In Dutch project, thermoplastic bio based material developed by Henkel was used. Nevertheless, Henkel is currently running some tests with a new developed eco-concrete that may be used in later stages of the Canal House project in order to increase compressive strength of printed pieces. For this phase of the project, building components, easy to join together, with gaps inside in a shape of honeycombs, were designed to be filled with special lightweight concrete assuring the insulation of the building, thanks to its air-entrained structure.[5] Every element consists of numerous diagonal hollow columns that will support entire structure. A house will have 13 rooms printed on site and be assembled into one house. Another advantage of the building is the fact that all parts can be also separated, in case house needs to be relocated.

In Chinese WinSun project stereolithography printing used a mix of industrial wastes, fibreglass, cement and hardening agent. Developed material allowed to create building components layer by layer, like in ordinary 3D technology. Desired mixture needs to have maximum workability as well as maximum flowability in order to be easily placed in layers. The layers must ensure the bonding with subsequent layers at the same time. As the compressive strength is required the water content should be minimized while appropriate flowability is maintained. The best describing word for the appropriate 3D cementitious mix would be thixotropic. The liquid state material should harden in appropriate time and before next layer is being laid. Engineers are working to find the best recipe for quick-setting concrete that will be manageable enough to be pumped out of the printer's nozzle and be as strong as reinforced concrete.

The possible material solution for 3D printing of building components could be sulphur concrete which is a composite material made of sulphur and aggregates (generally a coarse aggregate made of gravel or crushed rocks and a fine aggregate such as sand). The mix is heated above the melting point of sulphur ca. 140℃. After cooling the concrete reaches the target strength, without prolonged curing time like normal concrete. Sulphur concrete is considered as a potential building material for a lunar base shelter.

14.3 Application of 3D Printing Reproduction of Historical Building Ornamental Components

Construction is an ancient human activity; the structures created have engendered society's ability to function and prosper. Some of them are deemed to be of outstanding historical, aesthetic, or cultural importance and often provided with a special status (i.e. landmark designation) ordaining their conservation. Since the emergence of civilization, generally associated with the final stages of the Neolithic Revolution (9130 BCE), types of materials, methods and technologies used for construction have changed significantly. The materials used to construct buildings of historic significance are invariably subjected to an array of factors contributed to their decay and failure such as air pollution, salts, biodeterioration and mechanical loads (usage and traffic). Consequently, historical structures need to be conserved for future generations to connect with their ancestors and respect their achievements. Such conservation also allows people from other cultures to understand the values and beliefs that have shaped a civilization.

In the conservation work of a historical building, there is a standard of ethics that highlights the requirements of proposed interventions and encourages a minimum effective intervention. Seven essential degrees of interventions are made at various scales and levels of intensity according to the physical condition, causes of deterioration and prospective future environment of the cultural property under treatment. These interventions, which are likely to occur simultaneously in a major conservation project, are successively: (1) Prevention of deterioration; (2) Preservation; (3) Consolidation; (4) Restoration; (5) Rehabilitation; (6) Reproduction; and (7) Reconstruction. The seven degrees of conservation interventions are adopted in different application scenarios. The three interventions of restoration, reproduction and reconstruction involve manufacturing or replacing missing or damaged elements or components of historical buildings. A virtual reproduction process of a building element or component often serves as a restoration task. There are however two challenges that curb the reproduction process:

1) Drawings and other documentation are often missing or are scant, which renders it almost impossible to restore or reconstruct them to their original condition and incompatible conservation interventions without a forehand measurement can enhance the decay in historical buildings. Due to the idiosyncrasy of their construction and location, conventional measurement methods such as man-

ual physical mapping or "expert-naked-eye" analysis are laborious and inefficient, while some cultural properties are forbidden to directly touch or intervene for fear of erosion by a pollution of detection devices or human skins. This leads to a demand of an efficient digital non-contact or indirect measurement method.

2) Stone and masonry structures or components widely exist in historical buildings and cultural heritages. There is a propensity for structures and components of such materials to have detailed engravings, which may be on complex curved surfaces.[6] To manually reproduce or reconstruct these components is an arduous task; the number of artisans in China, for example, who have skills and experience to undertake this work is limited and even decreasing, and the cost to manufacture templates for their construction activities is an expensive undertaking. New adaptable automatic restoration methods are encouraged for architectural conservation work.

Yet with the advancement and development of digital technologies, such problems can be overcome. Three-dimensional (3D) scanning imaging systems and photogrammetric shape measurement systems are capable of measuring space dimension of existing structures and artefacts. As-built surface information can be acquired without physical contact, while digital photos and other camera or vision based systems can be used for monitoring, recognition, localization and tracking. Combined with 3D printing technology, which is an automatic manufacturing process without templates, the potential and increasing applications of 3D scanning for restoring or reconstructing historical heritage buildings for people to enjoy is unbounded. However, there is limited research that have used these technologies to restore key components of historical buildings. This paper presents a novel digital process for reproducing a whole ornamental component of a historical building using a combination of 3D laser scanning and cement mortar-based 3D printing technology. Cement mortar is used for that purpose, as is characterizes the physical properties of stone or rock and performs harmoniously with the original stony material in color, tone, texture, form and scale; this material conforms to the requirements embedded in conservation ethics.

14.3.1 Proposed Digital Reproduction Process

The proposed digital reproduction task for a historical building ornamental component is independently developed in a digital construction system that actualizes 3D scanning and cement mortar-based 3D printing. The process is divided into the following four steps and includes algorithms that have been developed for executing this task (Fig. 14-3):

(1) Model acquisition: The dimensional data of an intact historical building component is directly measured using a hand-held structured light 3D scanner which is a triangulation laser scanner and has a high accuracy and mobility for short-range (<1m focal distance) scanning; then the point cloud data (PCD) is inputted into a reverse engineering software (e.g., Geomagic Studio), which automatically encapsulate the PCD into a 3D solid model.

(2) Program generation: A STereo Lithography (STL) file is formatted from the solid model and inputted into the data processing module of the system. This is used to generate a machine control program in accordance with a hierarchical algorithm for model slicing and a modified scan line

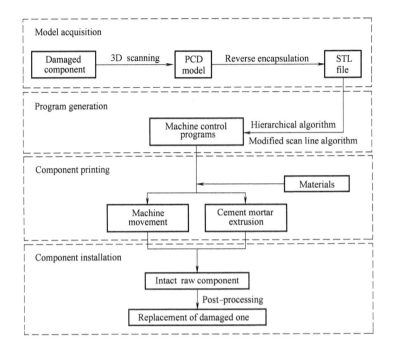

Fig. 14-3 The process flow chart.

algorithm for nozzle path planning. The development of these algorithms is presented below in the next section of this paper.

(3) Component printing: The control programs coordinate the machine movement and cement mortar extrusion so as to layer the construction of the component.

(4) Component installation: The printed component is post-processed (e. g., polishing) and installed to replace the damaged one in the historical building; this then finalizes the reproduction process.

The digital construction system contains five modules (Fig. 14-4): (1) modeling, (2) control, (3) motion, (4) extrusion, and (5) data processing.

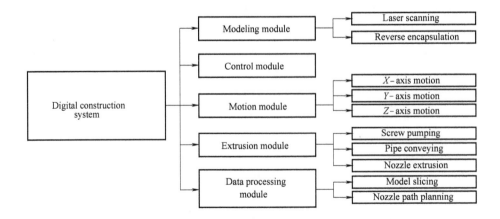

Fig. 14-4 System Composition.

14.3.2 Process Validation

A column is one of the most common and essential components of a building responsible for bearing loads, as well possessing aesthetic features. A plinth is usually used in a complete column structure for delivering upper load, water resistance, collision avoidance, lateral support and aesthetics. However, stone plinths used in historical Chinese timber buildings are often prone to deterioration; they are often of curved shapes or even heteromorphosis and no drawings exist; an example is presented in Fig. 14-5.

a)　　　　　　　　　　　　b)　　　　　　　　　　　　c)

Fig. 14-5　Damaged stone plinths of varying degrees.

a) Almost intact plinth　b) Partly damaged plinth　c) Completely damaged plinth

Within the campus at the Huazhong University of Science and Technology (HUST), there are also damaged stone historical plinths similar to those in Fig. 14-5 and one of them with simple curved surface has been selected for this research to demonstrate the feasibility of the proposed digital reproduction process. Notably, the damaged plinth functions as an ornamental component for a timber column structure. The role of the plinth is to protect the timber column from water damage and collision, as well provide lateral support and decorative aesthetics.[7]

According to previous jump table test, a fine aggregate fiber cement mortar is made up for reproduction of the plinth, which has a 0.3 water-cement ratio and a 1 : 1 sand : cement ratio, plus 0.1% micro polypropylene fibers, including 1.25% water reducer.

For the purpose of the reproduction process, a hand-held structure light 3D scanner, Creaform MetraSCAN 3D, is used to obtain the dimensional information of an intact plinth. Prior to scanning the plinth, a preparatory work is done as instrument calibration for the C-Track dual camera sensor, which is erected nearby to establish a spatial reference coordinate system for scanning measurement. The scanner is steadily held and revolves around the column, which is repeated several times. The high-speed laser measurement and data transmission realizes real-time surface rendering of the measured plinth as well as the column and surrounding environment on the workstation screen when the scanner is working.

Reverse engineering software known as Geomagic Studio processes the measurement signals with a PCD model, which is created and displayed on the screen. An optimized mesh generation treatment for the PCD model is conducted using the software after the redundant PCD of the column and surrounding environment is manually removed. This is an encapsulation and modification process to obtain an accurate and valid plinth solid STL model that can be used for printing. The 3D scanning and PCD treatment process is presented in Fig. 14-6.

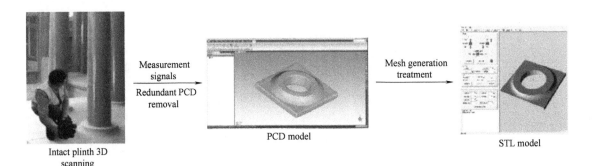

Fig. 14-6 3D scanning and PCD treatment process of the intact plinth.

The plinth printing process is dependent on the self-forming and stabilization of cement mortar layering. It takes approximately 10min to 15min to mix the materials, which is able to be cured within one hour. There is no discontinuity in the printing process, as a bucket of the mixture is poured into the feed hopper by manual as soon as the current mixture is almost completely pumped.[8] Before the nozzle deposits the material along the setting paths, water and neat cement paste are successively pumped and extruded from the nozzle's head, to lubricate the passage, particularly the pipeline.

The intact plinth solid model is divided into two parts before printing. Four similar half plinths have been printed. Two of the half plinths are intended for a compressive strength test. The other two afford the restoration task by joining together to form the entire plinth when installed.[9] The footprint of the entire plinth is 0.7m in diameter and the height is approx. 40 ~ 80mm. The plinth shape is akin to a curved cup with a circular hollow (diameter of approx. 370 mm) inside it and consists of four layers (Fig. 14-7). The average printing time of 2 min per layer was recorded.

The ribbed surface finish of the two printed half plinths need to be polished for an improved smooth surface reproduction effect by using a sander and abrasive paper, especially the surfaces in touch with the timber column and the ground. Then, they can be painted using imitation stone paint or undergo an old process treatment. The two printed half plinths enclose the column using a structural adhesive to bond them together. Notably, the diameter of the column is 350 mm so that 10 mm gap between the column and plinth is considered for infilling the structural adhesive. Residual part of the original damaged plinth is removed away before replacement and the completely reproduced plinth can be seen in Fig. 14-8.

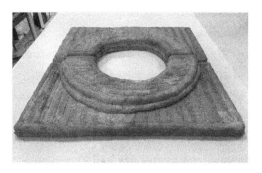

Fig. 14-7 The two printed half individual plinths.

Fig. 14-8 Installation of the printed plinth after post-processing.

14.3.3 Reproduction Evaluation

The digital reproduction process for the historical plinth presented is novel and has the potential to replace the manual procedure for a restoration task. However, it is necessary to evaluate this new approach process to ensure that the printed plinth provides a similar exterior as the original one and the required structural performance so as to be used in practice.

The scanning resolution is high for such a concrete building component, despite the encapsulation processes modeling is an approximation modeling of the physical intact plinth. The PCD file contains tens of thousands of points, which possesses an average distance of 0.10 mm between two points (Fig. 14-9a). The surface of the 3D solid model (STL model) is a triangular mesh incorporating a vast number of tiny triangular planes determined by any of three points (Fig. 14-9b). The digital representation surface and geometry is akin to the original physical artefact;[10] however, the core is slightly larger to accommodate the structural adhesive filling.

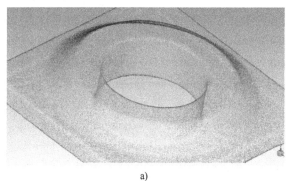

a)   b)

Fig. 14-9

a) Detail of the PCD model of the intact plinth after manual treatment

b) the 3D solid STL model of the intact plinth

In terms of visualization, there remains an obvious distinction between the printed and the original plinth. This is due to the printing art and the material as: (1) the layered stack art that results in a vertical stratified surface; (2) the nozzle extrusion process that results in the cascade façade;

(3) and the material proportion that affects the surface planeness of per strip or layer. In this instance resolution discrepancies between scanning and printing can materialize.

Factors such as size, shape, color, and uniformity, are used to describe the façade features of the printed plinth. A straight steel ruler and a band tape are taken to measure size parameters. The shape and color of the printed plinth can be observed by the naked eye. A steel ruler and a feeler are used to measure the strip intervals uniformity. The printed plinth is 0.7m wide in diameter of the bottom surface and the height is approx. 40~80mm with a curved cup shape and hoar façade, which is similar to the original intact one. Yet the reproduction quality of curved side wall of the cup-shape plinth is unsatisfactory. As the deposited strip, controlled by the nozzle diameter, was too thick to form the fine curved side wall, which is represented by a two-ladder surface. In addition the rough surface does not resemble the original due to its layered texture, and has a darker color tone compared to natural stone.

The distance of the central lines of adjacent strips (A) is approximately 17.0~21.5mm. The distance between two adjacent strips (interval) (B) is approximately 0.5~3.0mm and the width of a strip (C) is approximately 15.0~20.0mm. The concrete texture, the mild variation of width of strips and intervals, result from the non-perfect coordination of the pump pressure output and machine (nozzle) movement. This particularly arises for short filling paths that are thick at their endpoints and thinner at the middle sections. The 15mm wide nozzle diameter has a direct impact on the surface uniformity, which can be improved with more strips within one unit area when the nozzle diameter is diminished. Therefore, the post-processing work is necessary for a practical reproduction of a historical building component (Fig. 14-8).

14.3.4　Conclusion

The process of 3D scanning is becoming an important tool for conserving and reproducing cultural heritage artefacts. Such technology also provides an ability to acquire high spatial resolution data needed to ameliorate the effectiveness of the reproduction process. Concrete based 3D printing is now gaining interest and momentum as it begins to become mature; 3D scanning of a physical object provides a mechanism to derive virtual data and information and when integrated with 3D printing the reproduction of an artefact becomes a reality. The combined use of such technologies provides unbounded opportunities for stimulating innovation and a new digital construction process, with corresponding algorithms for reproducing a historical building ornamental component. The approach proposed in this paper is validated through an overall digital reproduction of an individual plinth of a damaged curved cup shape stone, which demonstrates the future of combined use of 3D scanning and 3D printing in solving a pervasive and on-going problem in architectural restoration work.

14.3.5　Limitations and Future Research

The research presented combined 3D scanning and 3D printing for historical building restoration, and demonstrated the potential of this approach to unlock a new restoration process. There remain some limitations, which need to be acknowledged to engender future work in this area:

1) Restoration of historical buildings often requires the use of the similar material. This trial uses cement mortar to substitute the stone in situ, while the technology and material science may have some room for improvement such as high-index cement mortar with steel fiber reinforcement for higher compressive strength. A mixture of sand and mountain flour with adhesive material such as epoxy resin in the future work will enhance the ability to produce a more befitting replica for a stone component.

2) Historical building components usually contain intricate carving patterns or designs on their surface, which require a high accuracy during reproduction. The components reproduced by this process are relatively crude and are unable to accommodate the need for reproducing details. Improved control of mechanical three axis motion and an intelligent nozzle for higher accuracy and flexibility of material deposition is necessary for this issue to be addressed.

3) The reproduction process demonstrated refers to overall replication of an ornamental element or component, which entails the 6th degree of conservation intervention. Restoration as replacement or spot priming for partial damages, corrosions or losses, as the 4th degree of conservation intervention, requires decayed material detection techniques, smart restoration modeling algorithms for substitutable or missing parts, modified printing arts as well as compatible installation techniques. This is more comprehensive and of greater significance. The research has provided the initial building blocks for integrating 3D scanning and 3D printing technologies. The platform that has been established along with the algorithms will provide researchers with the ability to replicate and improve the system that has been proposed. It is suggested that the real challenges ahead for construction pertain to the issues of how to reproduce the original natural materials that have been used in historical buildings.

■ 14.4 Preparation of Computer Models for 3D Printing

So far we have discussed issues related to physical part of the manufacturing process. These are of course the key issues for the adoption of 3D printing technology. An important element is however also preparation of computer model for the parts to be manufactured. Fortunately, the level of 3D computer graphics both in terms of software and hardware makes it possible to build such digital models without much difficulties. It can be done using many commercial as well as Open Source software packages.

14.4.1 STL Data Format

Transferring model data via STL format requires constructing a triangulation of all boundary surfaces as illustrated in Fig. 14-10. This is most easily done if the solid model is build using B-Rep (boundary) representation, as for this representation the solid boundaries are stored explicitly within the model. The key element of exporting such representation to STL format is triangulation of curved surfaces. For CSG (Computed Solid Geometry) models, in order to save them in STL format, additional processing steps are needed to recover model boundaries.

Fig. 14-10

a) Honeycomb   b) Honeycomb mesh

While STL is a popular input format for 3D printing, it should be stressed that saving model in this format in not mandatory and the commands for controlling a printing process can be generated directly from model in its native representation. It is just the question of geometric computations within a geometric kernel of 3D printer. The current processing and data formats for geometric models for 3D printing have their roots in 3D computer graphics used for screen rendering. This can result in some difficulties and restrictions in using digital models for printing as discussed in the next point. It is worth to mention, that as additive manufacturing technologies mature and allow to produce very sophisticated, non-standard geometrical forms, there is also active research concerned new geometric representations that are related directly to the physical material deposition process. This allows to bypass the geometry processing that is suitable for screen rendering but problematic in case of 3D printing.

14.4.2 Preparing 3D Models

As said above, in most cases in order to send digital model for printing it is enough to save it in STL format. Many 3D computer graphics programs can export models in STL. However, one should be careful using them, because many of these programs are designed to be used primarily for screen rendering of 3D models. It means that they can tolerate specific features of the models that are non-essential for rendering but that will be crucial for 3D printing. The main points to pay attention to are:

1) 3D printing is a physical process contrary to screen rendering. Thus one has to obey physical constraints. Designing a model for printing one has to ensure that all elements of the model are physically realizable. This means for instance that free 1D edges and 2D faces are not allowed in the model.

2) Printing is done in the presence of gravity. One has to consider stability of the model and the weight of its parts to avoid damaging printed parts, for instance by breaking to slender support elements.

3) Some printing technologies require to design holes through which excess of non-bounded material can be evacuated.

4) The boundary surfaces of the model must be watertight, that means all faces must be connected and have consistent orientation of surface normals. This is in order to distinguish in unique way model interior and exterior space.

5) The triangulated surfaces must form a 2D manifold. In particular, all edges must be shared by exactly two faces, and there should be no singular points, where the boundary of the model touches itself. Example of non-manifold model is illustrated in Fig. 14-11.

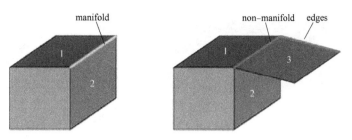

Fig. 14-11 Illustration of manifold and non-manifold edges of a 3D model.

14.4.3 Software

Autodesk Inventor is a software that allows to build complete 3D model of designed construction or device and enable to create planar drawing documentation for the project. While using Inventor, most of the time that constructor needs to put in a project is sacrificed for creative and conception works. All the changes made in a model are automatically transferred to the drawings.

In our investigations of digital side of 3D printing technology, we primarily use Autodesk Inventor and Blender. Blender is open source package for 3D modeling, animations and computer games production. Interesting feature of Blender is the export/import module for IFC models based on IfcOpenShell library. This module allows import and further processing of models prepared in BIM applications such as Revit or Tekla.

Two 3D models of building envelope components have been designed in Autodesk Inventor software. Replicas of the wall from the Canal House and WinSun houses were prepared and printed using ABS material (Acrylonitrile Butadiene Styrene) and using standard RepRap 3D printer (Fig. 14-12, Fig. 14-13).

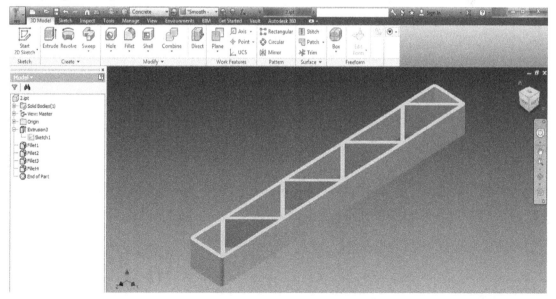

Fig. 14-12 3D model of an envelope component developed by WinSun company and Autodesk Inventor software environment.

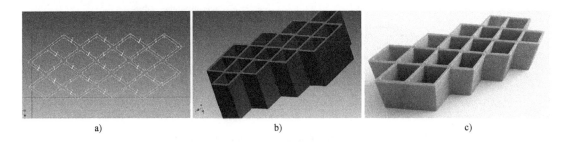

Fig. 14-13

a) 2D sketch　b) 3D model　c) Printed replica of a Canal House wall

New Words and Expressions

stereolithography　*n.*　立体平版印刷

rapid prototyping　快速成型

synthetic resins　合成树脂

depolymerization process　解聚过程

laser　*n.*　激光

nylon　*n.*　尼龙

polystyrene　*n.*　聚苯乙烯

ductile materials　*n.*　延性材料

extrude　*v.*　喷出

nozzle　*n.*　喷嘴

thermoplastic materials　热塑性材料

biodegradable　*adj.*　可生物降解的

fabrication　*n.*　制造，建造，装配

adhesive　*adj.*　带黏性的；*n.*　黏合剂

crosswise　*adj.*　交叉的

sulphur concrete　硫黄混凝土

idiosyncrasy　*n.*　特质

engraving　*v.*　雕刻

photogrammetric　*adj.*　摄影测量的

artefact　*n.*　人工制品

algorithm　*n.*　算法

heteromorphosis　*n.*　形态变异，异形

Notes

1. First step in creating the technology was invention of additions to the synthetic resins that af-

Unit 14　Introduction to 3D Printing of Buildings and Building Components

ter lightening of the resins, were causing start of depolymerization process.

本句难点解析：联系上下文可知，这里的 technology 是指的 stereolithography，即立体平版印刷。lighten 最常用的意思是减轻、发亮，但是在本句中与 resin（树脂）一起使用，解释为漂白。

本句大意如下：创建这项技术的第一步是合成树脂添加剂的发明，这些添加剂能在漂白树脂后，引起解聚过程的开始。

2. Ductile materials which are hardening itself during cooling process, are extruded through double headed nozzle.

本句难点解析：本句 which 后面都是用于修饰 ductile materials，extrude 解释为"喷出、压出"，harden 有"坚固，硬化"之意，在此处为"淬火"。

本句大意如下：在冷却过程中淬火的延性物料通过双头喷嘴喷出。

3. Both, modelling and supportive materials are being deposited according to the cross-section layers, generated from digital model supporting the printer.

本句难点解析：本句主干为 modelling and supportive materials are being deposited，后面的 according to 是对中心句的补充解释，generated from 及其后是对 cross-section layers 的解释，supporting 及其后是对 digital model 的解释。deposite 解释为"沉积"，也就是说在打印的过程中，材料喷涂在底板上，可以认为这也是一种"沉积"。

本句大意如下：模型和支撑材料会按照截面图层进行喷涂，这种截面图层是由支持打印机的数字模型生成的。

4. The idea of the inventor is to create a printer, that will have one or few nozzles that are moving on two parallel lanes installed at the construction site, separated from themselves a few meters wider than the width of the building.

本句难点解析：第一个 that 后内容是为了修饰 printer，第二个 that 后内容是为了修饰 nozzles。

本句大意如下：发明人的想法是创建一种有一个或几个喷嘴的打印机，这些喷嘴可在施工现场安装的两个平行轨道上移动，且彼此间距比楼房宽度宽几米。

5. For this phase of the project, building components, easy to join together, with gaps inside in a shape of honeycombs, were designed to be filled with special lightweight concrete assuring the insulation of the building, thanks to its air-entrained structure.

本句难点解析：本句主干为 building components were designed to be filled with special lightweight concrete，其他的部分都是修饰成分。

本句大意如下：在项目的这个阶段，容易组装、内部有蜂窝形状空隙的建筑构件，将根据设计，由特殊的轻质混凝土进行填充。由于其具有夹带空气的结构，可以确保建筑物的保温性能。

6. There is a propensity for structures and components of such materials to have detailed engravings, which may be on complex curved surfaces.

本句难点解析：propensity 意为倾向。

本句大意如下：采用这种材料的结构和构件很容易进行精细雕刻，该雕刻可能是在复杂的曲面上。

7. The role of the plinth is to protect the timber column from water damage and collision, as well provide lateral support and decorative aesthetics.

本句难点解析：本句中，plinth 意为"底座"。文中提到为木柱制作一个底座，这是考虑到木材易受腐蚀的特点，也是底座的一个主要作用。

本句大意如下：底座的作用是保护木柱免受水的破坏和碰撞破坏，并提供侧面支撑和装饰美感。

8. There is no discontinuity in the printing process, as a bucket of the mixture is poured into the feed hopper by manual as soon as the current mixture is almost completely pumped.

本句难点解析：句中 feed hopper 意为"进料斗"。

本句大意如下：在印刷过程中没有间断，因为一旦混合物被完全泵送，另一整桶混合物就又被手动地注入料斗之中。

9. The other two afford the restoration task by joining together to form the entire plinth when installed.

本句难点解析：根据上下文，可以知道 the other two 指的是 plinths（底座），afford 原意"提供，买得起"，在这里与 task 相连，可以解释为"承担"。

本句大意如下：另外两个底座通过在安装时连接在一起形成整个底座，来承担修复工作。

10. The digital representation surface and geometry is akin to the original physical artefact.

本句难点解析：句中 be akin to 意为"类似地"。

本句大意如下：数字表示面和几何形状类似于原始的人工制品。

Exercises

Translate the following phrases into Chinese.

1. 3D printer

2. innovative application

3. appropriate liquid state

4. prefabricated element

5. Contour Crafting (CC) technology

6. unlimited architectural flexibility

7. prevention of deterioration

8. expert-naked-eye analysis

9. triangulated surface

10. Autodesk Inventor and Blender

Translate the following Sentences into Chinese.

1. The idea of 3D printing was born already in 1983, when Charles W. Hull came up with an idea of hardening the tabletop coatings with the UV light.

2. Architects from Dus Architects want to prove, that by printing components of the house directly on the site, they will be able to completely eliminate building waste and minimize costs of the

transport.

3. The scanning resolution is high for such a concrete building component, despite the encapsulation processes modeling is an approximation modeling of the physical intact plinth.

4. Restoration of historical buildings often requires the use of the similar material. This trial uses cement mortar to substitute the stone in situ, while the technology and material science may have some room for improvement such as high-index cement mortar with steel fiber reinforcement for higher compressive strength.

5. An important element is however also preparation of computer model for the parts to be manufactured. Fortunately, the level of 3D computer graphics both in terms of software and hardware makes it possible to build such digital models without much difficulties.

Unit 15

BIM: Acceptance Model in Construction Organizations

15.1 Introduction to BIM

The architecture, engineering, and construction (AEC) industry has long sought techniques to decrease project cost, increase productivity and quality, and reduce project delivery time. Building information modeling (BIM) offers the potential to achieve these objectives. BIM simulates the construction project in a virtual environment. With BIM technology, an accurate virtual model of a building, known as a building information model, is digitally constructed. When completed, the building information model contains precise geometry and relevant data needed to support the design, procurement, fabrication, and construction activities required to realize the building. After completion, this model can be used for operations and maintenance purposes. Fig. 15-1 depicts the typical applications of BIM at different stages of the project life cycle.

A building information model characterizes the geometry, spatial relationships, geographic information, quantities and properties of building elements, cost estimates, material inventories, and project schedule. The model can be used to demonstrate the entire building life cycle. As a result, quantities and shared properties of materials can be readily extracted. Scopes of work can be easily isolated and defined. Systems, assemblies, and sequences can be shown in a relative scale within the entire facility or group of facilities. Construction documents such as drawings, procurement details, submittal processes, and other specifications can be easily interrelated.

BIM can be viewed as a virtual process that encompasses all aspects, disciplines, and systems of a facility within a single, virtual model, allowing all design team members (owners, architects, engineers, contractors, subcontractors, and suppliers) to collaborate more accurately and efficiently than using traditional processes.[1] As the model is being created, team members are constantly refining and adjusting their portions according to project specifications and design changes to ensure the model is as accurate as possible before the project physically breaks ground.[2]

It is important to note that BIM is not just software; it is a process and software. BIM means not only using three-dimensional intelligent models but also making significant changes in the workflow and project delivery processes. BIM represents a new paradigm within AEC, one that encourages integration of the roles of all stakeholders on a project. It has the potential to promote greater efficiency

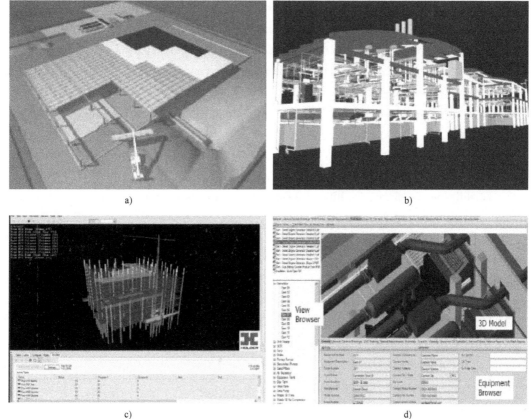

Figure. 15-1 Different components of a building information model:
MEP = mechanical, electrical, and plumbing.

a) Site Logistic Planning Model b) Integrated Structural & MEP Model c) Construction Sequencing Model
d) Facility Information Model

and harmony among players who, in the past, saw themselves as adversaries. BIM also supports the concept of integrated project delivery, which is a novel project delivery approach to integrate people, systems, and business structures and practices into a collaborative process to reduce waste and optimize efficiency through all phases of the project life cycle.

15.2　Applications of Building Information Modeling

A building information model can be used for the following purposes:

1) Visualization: 3D renderings can be easily generated in house with little additional effort.

2) Fabrication/shop drawings: It is easy to generate shop drawings for various building systems. For example, the sheet metal ductwork shop drawings can be quickly produced once the model is complete.

3) Code reviews: Fire departments and other officials may use these models for their reviews of building projects.

4) Cost estimating: BIM software has built-in cost estimating features. Material quantities are

automatically extracted and updated when any changes are made in the model.

5) Construction sequencing: A building information model can be effectively used to coordinate material ordering, fabrication, and delivery schedules for all building components.

6) Conflict, interference, and collision detection: Because building information models are created to scale in 3D space, all major systems can be instantly and automatically checked for interferences. For example, this process can verify that piping does not intersect with steel beams, ducts, or walls.

7) Forensic analysis: A building information model can be easily adapted to graphically illustrate potential failures, leaks, evacuation plans, and so forth.

8) Facilities management: Facilities management departments can use it for renovations, space planning, and maintenance operations.

The key benefit of a building information model is its accurate geometrical representation of the parts of a building in an integrated data environment. Other related benefits are as follows:

1) Faster and more effective processes: Information is more easily shared and can be value-added and reused.

2) Better design: Building proposals can be rigorously analyzed, simulations performed quickly, and performance benchmarked, enabling improved and innovative solutions.

3) Controlled whole-life costs and environmental data: Environmental performance is more predictable, and lifecycle costs are better understood.

4) Better production quality: Documentation output is flexible and exploits automation.

5) Automated assembly: Digital product data can be exploited in downstream processes and used for manufacturing and assembly of structural systems.

6) Better customer service: Proposals are better understood through accurate visualization.

7) Lifecycle data: Requirements, design, construction, and operational information can be used in facilities management.

After gathering data on 32 major projects, Stanford University's Center for Integrated Facilities Engineering reported the following benefits of BIM:

1) Up to 40% elimination of unbudgeted change.

2) Cost estimation accuracy within 3% as compared to traditional estimates.

3) Up to 80% reduction in time taken to generate a cost estimate.

4) A savings of up to 10% of the contract value through clash detections.

5) Up to 7% reduction in project time.

15.3 Role of BIM in the AEC Industry: Current and Future Trends

In this section, the role of BIM in the AEC industry and its current and future trends are discussed based on the results of two questionnaire surveys. McGraw-Hill Construction published a comprehensive market report of BIM's use in the AEC industry in 2008 and projections for 2009 based on the findings of a questionnaire survey completed by 82 architects, 101 engineers, 80 contractors, and 39 owners (total sample size of 302) in the United States. Some of the key findings are as follows:

1) Architects were the heaviest users of BIM—43% used it on more than 60% of their projects-while contractors were the lightest users, with nearly half (45%) using it on less than 15% of projects and only a quarter (23%) using it on more than 60% of projects.

2) Eighty-two percent of BIM users believed that BIM had a very positive impact on their company's productivity.

3) Seventy-nine percent of BIM users indicated that the use of BIM improved project outcomes, such as fewer requests for information (RFI) and decreased field coordination problems.

4) Sixty-six percent of those surveyed believed use of BIM increased their chances of winning projects.

5) Two-third of users mentioned that BIM had at least a moderate impact on their external project practices.

The report predicted that prefabrication capabilities of BIM would be widely used to reduce costs and improve the quality of work put in place. As a whole, BIM adoption was expected to expand within firms and across the AEC industry.

Kunz and Gilligan conducted a questionnaire survey to determine the value from BIM use and factors that contribute to success. The main findings of their study are as follows:

1) The use of BIM had significantly increased across all phases of design and construction during the past year.

2) BIM users represented all segments of the design and construction industry, and they operated throughout the United States.

3) The major application areas of BIM were construction document development, conceptual design support, and preproject planning services.

4) The use of BIM lowered overall risk distributed with a similar contract structure.

5) At the time of the survey, most companies used BIM for 3D and 4D clash detections and for planning and visualization services.

6) The use of BIM led to increased productivity, better engagement of project staff, and reduced contingencies.

7) A shortage was noted of competent building information modelers in the construction industry, and demand was expected to grow exponentially with time.

The results of these surveys indicate that the AEC industry still relies very much on traditional drawings and practices for conducting its business. At the same time, AEC professionals are realizing the power of BIM for more efficient and intelligent modeling. Most of the companies using BIM report in strong favor of this technology. The survey findings indicate that users want a BIM application that not only leverages the powerful documentation and visualization capabilities of a CAD platform but also supports multiple design and management operations. <u>BIM as a technology is still in its formative stage, and solutions in the market are continuing to evolve as they respond to users' specific needs.</u>[3]

15.4 BIM Benefits: Case Studies

In the above-mentioned surveys, the AEC industry participants indicated that BIM use resulted in time and cost savings. However, no data were provided to quantify and support these facts. The following four case studies illustrate the cost and time savings realized in developing and using a building information model for the project planning, design, preconstruction, and construction phases. All the data reported in this section were collected from the Holder Construction Company (HCC), a midsize general contracting company based in Atlanta, Georgia (hereinafter referred to as the general contractor, or GC).

15.4.1 Case Study 1: Aquarium Hilton Garden Inn, Atlanta, Georgia

The Aquarium Hilton Garden Inn project comprised a mixed-use hotel, retail shops, and a parking deck. Brief project details are as follows:

1) Project scope: $46 million, 484000-square-foot hotel and parking structure.
2) Delivery method: Construction manager at-risk (CM at-risk).
3) Contract type: Guaranteed maximum price.
4) BIM scope: Design coordination, clash detection, and work sequencing.
5) BIM cost to project: $90000, or 0.2% of project budget ($40000 paid by owner).
6) Cost benefit: Over $200000 attributed to elimination of clashes.
7) Schedule benefit: 1143 hours saved.

Although the project had not been initially designed using BIM technology, beginning in the design development phase, the GC led the project team to develop architectural; structural; and mechanical, electrical, and plumbing models of the proposed facility, as shown in Fig. 15-2. These models were created using detail-level information from subcontractors based on drawings from the designers.[4]

After the initial visualization uses, the GC began to use these models for clash detection analysis. This BIM application enabled the GC to identify potential collisions or clashes between various structural and mechanical systems. During the design development phase, 55 clashes were identified, which resulted in a cost avoidance of $124500. Just this stage alone yielded a net savings of $34500 based on the original building information model development cost of $90000. At the construction documents phase, the model was updated and resolved collisions were tracked. Each critical clash was shared with the design team via the model viewer and a numbered collision log with a record of individual images of each collision per the architectural or structural discipline.[5] The collision cost savings values were based on estimates for making design changes or field modifications had the collision not been detected earlier. More than 590 clashes were detected before actual construction began. The overall cost savings based on the 590 collisions detected throughout the project was estimated at $801565, as shown in Table 15-1. For calculating net cost savings, a conservative approach was adopted by assuming that 75% of the identified collisions can be detected through conventional practices (e.g., sequential composite overlay process using light tables) before actual construction begins. Thus, the net adjusted cost savings was roughly considered to be $200392.

Unit 15 BIM: Acceptance Model in Construction Organizations

Fig. 15-2 Building information models of the Aquarium Hilton Garden Inn Project
(Courtesy of Holder Construction Company, Atlanta, GA).
a) Architectural Model   b) Structural Model   c) Plumbing Model

Table 15-1 An illustration of Cost and Time Savings via Building Information Modeling in the Aquarium Hilton Garden Inn Project

Collision phase	Collisions	Estimated cost avoidance	Estimated crew hours	Coordination date
100% design development conflicts	55	$124500	—	30-Jun-2006
Construction (MEP collisions)				
Basement	41	$21211	50h	28-Mar-2007
Level 1	51	$34714	79h	3-Apr-2007
Level 2	49	$23250	57h	3-Apr-2007
Level 3	72	$40187	86h	12-Apr-2007
Level 4	28	$35276	68h	14-May-2007
Level 5	42	$43351	88h	29-May-2007
Level 6	70	$57735	112h	19-Jun-2007
Level 7	83	$78898	162h	12-Apr-2007
Level 8	29	$37397	74h	3-Jul-2007
Level 9	30	$37397	74h	3-Jul-2007
Level 10	31	$33546	67h	5-Jul-2007
Level 11	30	$45144	75h	5-Jul-2007
Level 12	28	$36589	72h	5-Jul-2007
Level 13	34	$38557	77h	13-Jul-2007
Level 14	1	$484	1h	13-Jul-2007
Level 15	1	$484	1h	13-Jul-2007
Subtotal construction labor	590	$564220	1143h	
20% MEP material value		$112844		
Subtotal cost avoidance		$801565		
Deduct 75% assumed resolved via conventional methods		($601173)		
Net adjusted direct cost avoidance		$200392		

Source: Holder Constuction Company, Atlanta, GA.

Note: MEP = mechanical, electrical, and plumbing.

During the construction phase, subcontractors also made use of these models for various installations. Finally, the GC's commitment to updating the model to reflect as-built conditions provided the owner a digital 3D model of the building and its various systems to help aid operation and maintenance procedures down the road.[6]

In a nutshell, the Aquarium Hilton Garden Inn project realized some excellent benefits through the use of BIM technology and certainly exceeded the expectations of the owner and other project team members. The cost benefits to the owner were significant, and the unknown costs that were avoided through collaboration, visualization, understanding, and identification of conflicts early were in addition to the reported savings.[7] After this project, the architect and GC began to use BIM technology on all major projects, and the owner used the developed building information model for sales and marketing presentations.

15.4.2 Case Study 2: Savannah State University, Savannah, Georgia

This case study illustrates the use of BIM at the project planning phase to perform options analysis (value analysis) for selecting the most economical and workable building layout. The project details are as follows:

1) Project: Higher education facility, Savannah State University, Savannah, Georgia.
2) Cost: $12 million.
3) Delivery method: CM at-risk, guaranteed maximum price.
4) BIM scope: Planning, value analysis.
5) BIM cost to project: $5000.
6) Cost benefit: $1995000.

For this project, the GC coordinated with the architect and the owner at the predesign phase to prepare building information models of three different design options. For each option, the BIM-based cost estimates were also prepared using three different cost scenarios (budgeted, midrange, and high range), as shown in Fig. 15-3. The owner was able to walk through all the virtual models to decide the best option that fit his requirements. Several collaborative 3D viewing sessions were arranged for this purpose. These collaborative viewing sessions also improved communications and trust between stakeholders and enabled rapid decision making early in the process. The entire process took 2 weeks, and the owner achieved roughly $1995000 cost savings at the predesign stage by selecting the most economical design option. Although it could be argued that the owner may have reached the same conclusion using traditional drawings, the use of BIM technology helped him make a quick, definitive, and well-informed decision.

15.4.3 Case Study 3: The Mansion on Peachtree, Atlanta, Georgia

The Mansion on Peachtree is a five-star mixed-use hotel in Atlanta, Georgia. The project details are as follows:

1) Cost: $111 million.
2) Schedule: 29 months (construction).

Option/Aspect	Specifications	Option A	Option B	Option C
Front Elevation		Option A 	Option B 	Option C 
Plan				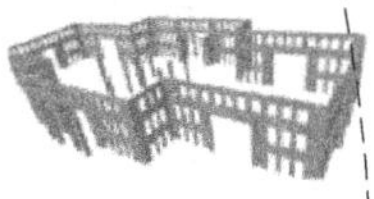
Stories	Not specified	2	2	3
Construction Funding	$11000000			
Max.Cost/GSF	$147.74			
Area/ft²		74459	87296	83018	73852
Net Area/ft²		46537	49125	50612	43338
Net to Gross Ratio		63%	56%	61%	59%
Cost Scenarios					
Budget: $147.74/ft²	$11000000	$12897111	$12270919	$10910894
Mid-Range: $175.00/ft²	$13030325	$15276800	$14535140	$12924100
High-Range: $200.00/ft²	$14891800	$17459200	$16611600	$14770400
Building Skin					
Primary Materials	Brick/Precast/Glass	Brick/Precast/Glass	Brick/Precast/Glass	Brick/Precast/Glass
Skin Articulation	Articulated, Trim	Articulated, Trim	Articulated, Trim	Articulated, Trim
Floor to Floor Height	—	14ft@1; 14ft@Upper	14ft@1; 14ft@Upper	14ft@1; 14ft@2; 12ft@Upper
Skin to Floor Ratio	—	58%	50%	39%
% Glass, % Brick	—	20% Glass, 80% Brick	28% Glass, 72% Brick	36% Glass, 64% Brick

Fig. 15-3 **Scope and budget options for the Savannah State Academic Building**

(Courtesy of Holder Construction Company, Atlanta, GA).

3) Delivery method: CM at-risk, guaranteed maximum price.

4) BIM scope: Planning, construction documentation.

5) BIM cost to project: $1440.

6) Cost benefit: $15000.

It was a fast-track project, and the GC identified the following issues at the project planning phase:

1) Incomplete design and documents.

2) Multiple uncoordinated consultants.

3) Field construction ahead of design.

4) Constant design development.

5) Owner's frequent scope changes.

The biggest challenge was how to maintain schedule and ensure quality with incomplete and uncoordinated design and how to minimize risk and rework. The project team decided to use BIM for project planning and coordination. First, contract documents were analyzed to flush out discrepancies and identify missing items. Then coordinated shop drawings were prepared via model extractions. These shop drawings were reviewed with the design team to resolve any conflicts and issue a field use set to subcontractors for coordination and construction.

Initially, the project designers presented two finishing options (brick vs. precast) to the owner, as shown in Fig. 15-4a. Via BIM viewer software, the owner visually compared both options and selected the precast one based on appearance and cost. Then, based on the project drawings, the GC prepared the 3D interior elevations to clarify interior details, as illustrated in Fig. 15-4b. If any component was found missing or conflicting with the other components, a RFI was issued to the designer to resolve this conflict before construction. Finally, a 4D scheduling model was prepared (Fig. 15-4c) to decide the construction sequence and align all resources. Through these measures, the project team was able to complete the project on time and within budget.

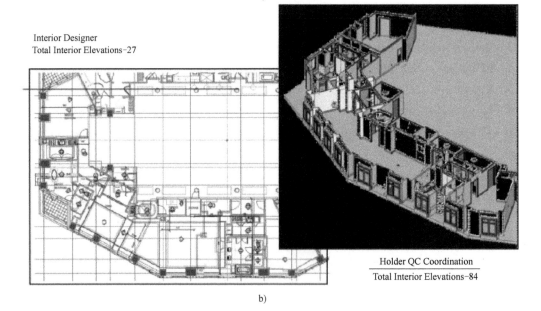

Fig. 15-4 Use of BIM in the Mansion on Peachtree Project

(Courtesy of Holder Construction Company, Atlanta, GA).

a) 3D Value Analysis for Visual Clarification b) Quality Control Coordination Shop Drawings and the Model

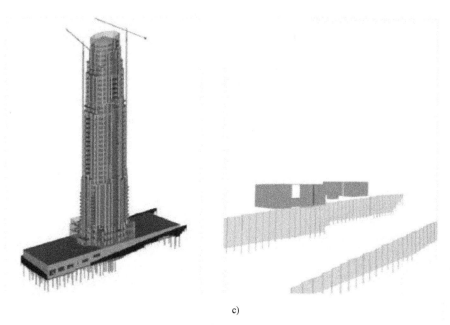

c)

Fig. 15-4 Use of BIM in the Mansion on Peachtree Project

(Courtesy of Holder Construction Company, Atlanta, GA). (Continued)

c) 4D Phasing Model

15.4.4 Case Study 4: Emory Psychology Building, Atlanta, GA

The Emory Psychology Building is a LEED-certified, 110000-square-foot facility on the campus of Emory University in Atlanta, Georgia. It is a multipurpose structure designed to provide instructional and research space. The project details are as follows:

1) Cost: $35 million.
2) Schedule: 16 months.
3) Delivery method: CM at-risk, guaranteed maximum price.
4) BIM scope: sustainability analyses.
5) BIM cost to project and cost benefit: n/a.

The project architect developed the building information model of the facility at the early design phase to determine the best building orientation and evaluate various skin options such as masonry, curtain wall, and window styles, as shown in Fig. 15-5. The building information model was also used to perform daylight studies, which, in effect, helped to decide the final positioning of the building on the site. To achieve this, views of the facility were established within BIM software using the software's sun positioning feature. Subsequently, shading and lighting studies and right-to-light studies were conducted to determine the effects of the sun throughout the year and the effects of the facility on surrounding buildings. Right-to-light studies were also conducted to evaluate lighting conditions at the proposed facility's courtyard space and those spaces adjacent to the courtyard.

As a direct result of these studies, the building's design was adjusted as follows:

1) Window openings on the west façade were reduced.

2) The penthouse, which is located on the roof of the building, was reduced in overall square footage.

3) The overall height of the building was reduced.

As all of these design adjustments were able to be incorporated during the design phase, the analyses prevented costly and time-consuming redesign at later stages in the project life cycle.

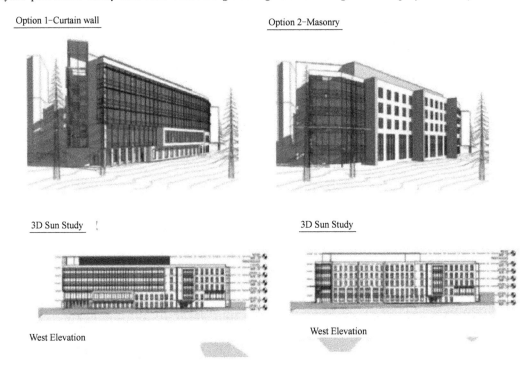

Fig. 15-5 Use of BIM for options analysis and sun studies in the Emory Psychology Building
(Courtesy of Holder Construction Company, Atlanta, GA).

15.5 Return on Investment Analysis

The return on investment (ROI) analysis is one of the many ways to evaluate a proposed investment. It compares the gain anticipated (or achieved) from an investment against the cost of the investment (i.e., ROI = earning/cost). ROI is typically used to evaluate many types of corporate investments, from research and development projects to training programs to fixed asset purchases.

The McGraw-Hill Construction survey of AEC industry participants indicated that 48% of respondents were tracking BIM ROI at a moderate level or above. It also found that the initial system cost did not seem to be a problem. Doubling the system cost could reduce ROI only by up to 20%. For this study, detailed cost data from 10 projects were acquired from HCC to perform the BIM ROI analysis. The results are shown in Table 15-2.

As evident from Table 15-2, the BIM ROI for different projects varied from 140% to 39900%. On average, it was 1633% for all projects and 634% for projects without a planning or value analysis phase. Because of the large data spread, it is hard to conclude a specific range for BIM ROI. The probable reason for this spread is the varying scope of BIM in different projects. In some projects, BIM savings were measured using "real" construction phase "direct" collision detection cost avoidance, and in other projects, savings were computed using "planning" or "value analysis" phase cost avoidance. Also, none of these cost figures account for indirect, design, construction, or owner administrative or other "second wave" cost savings that were realized as a result of BIM implementation. Hence, the actual BIM ROI can be far greater than reported here.

Table 15-2 Building Information Modeling Return on Investment Analysis

Year	Cost/ $M	Project	BIM scope	BIM cost/ $	Direct BIM savings/ $	Net BIM savings/ $	BIM ROI(%)
2005	30	Ashley Overlook	P/PC/CD	5000	(135000)	(130000)	2600
2006	54	Progressive Data Center	F/CD/FM	120000	(395000)	(232000)	140
2006	47	Raleigh Marriott	P/PC/VA	4288	(500000)	(495712)	11560
2006	16	GSU Library	P/PC/CD	10000	(74120)	(64120)	640
2006	88	Mansion on Peachtree	P/CD	1440	(15000)	(6850)	940
2007	47	Aquarium Hilton	F/D/PC/CD	90000	(800000)	(710000)	780
2007	58	1515 Wynkoop	P/D/VA	3800	(200000)	(196200)	5160
2007	82	HP Data Center	F/D/CD	20000	(67500)	(47500)	240
2007	14	Savannah State	F/D/PC/VA/CD	5000	(2000000)	(1995000)	39900
2007	32	NAU Sciences Lab	P/CD	1000	(330000)	(329000)	32900
Total all types				260528	4516620	4256092	1633
Totals without planning/VA phase				247440	1816620	1569180	634

Source: Holder Construction Company, Atlanta, GA.

Note: CD = construction documentation; D = design; F = feasibility analysis; FM = facilities management; GSU = Georgia State University; NAU = Northern Arizona University; P = planning; PC = preconstruction services; ROI = return on investment; VA = value analysis.

15.6 BIM Risks

BIM risks can be divided into two broad categories: legal (or contractual) and technical. In the following paragraphs, key risks in each category are briefly discussed.

The first risk is the lack of determination of ownership of the BIM data and the need to protect it through copyright laws and other legal channels. For example, if the owner is paying for the design, then the owner may feel entitled to own it, but if team members are providing proprietary information for use on the project, their proprietary information needs to be protected as well. Thus, there is no simple answer to the question of data ownership; it requires a unique response for every project depending on the participants' needs. The goal is to avoid inhibitions or disincentives that discourage participants from fully realizing the model's potential. To prevent disagreement over copyright issues, the best solution is to set forth in the contract documents ownership rights and responsibilities.

When project team members other than the owner and architect/engineer contribute data that are integrated into the building information model, licensing issues can arise. For example, equipment and material vendors offer designs associated with their products for the convenience of the lead designer in hopes of inducing the designer to specify the vendor's equipment.[8] While this practice might be good for business, the licensing issues can arise if the designs are not produced by a designer licensed in the location of the project.

Another contractual issue to address is who will control the entry of data into the model and be responsible for any inaccuracies. Taking responsibility for updating building information model data and ensuring its accuracy entails a great deal of risk. Requests for complicated indemnities by BIM users and the offer of limited warranties and disclaimers of liability by designers are essential negotiation points that need to be resolved before BIM technology is used. It also requires more time spent inputting and reviewing BIM data, which is a new cost in the design and project administration process. Although these new costs may be dramatically offset by efficiency and schedule gains, they are still a cost that someone on the project team will incur. Thus, before BIM technology can be fully used, not only must the risks of its use be identified and allocated, but the cost of its implementation must be paid for as well.

The integrated concept of BIM blurs the level of responsibility so much that risk and liability are likely to be enhanced. Consider the scenario in which the owner of the building files suit over a perceived design error. The architect, engineers, and other contributors to the BIM process look to each other in an effort to try to determine who had responsibility for the raised matter. If disagreement ensues, the lead professional not only will be responsible as a matter of law to the claimant but may have difficulty proving fault with others such as the engineers.

As the dimensions of cost and schedule are layered onto the building information model, responsibility for the proper technological interface among various programs becomes an issue. Many sophisticated contracting teams require subcontractors to submit detailed critical path method schedules and cost breakdowns itemized by line items of work prior to the start of the project. The general contractor then compiles the data, creating a master schedule and cost breakdown for the entire project. When the subcontractors and prime contractor use the same software, the integration can be fluid. In cases where the data are incomplete or are submitted in a variety of scheduling and costing programs, a team member—usually a general contractor or construction manager-must re-enter and update a master scheduling and costing program.[9] That program may be a BIM module or another program that is integrated with the building information model. At present, most of these project management tools have been developed in isolation. Responsibility for the accuracy and coordination of cost and scheduling data must be contractually addressed.

One of the most effective ways to deal with these risks is to have collaborative, integrated project delivery contracts in which the risks of using BIM are shared among the project participants along with the rewards. Recently, the American Institute of Architects released an exhibit on BIM to help project participants define their BIM development plan for integrated project delivery. This exhibit may assist project participants in defining model management arrangements, as well as authorship, ownership, and level-of-development requirements, at various project phases.

15.7 BIM Future Challenges

The productivity and economic benefits of BIM to the AEC industry are widely acknowledged and increasingly well understood. Further, the technology to implement BIM is readily available and rapidly maturing. Yet BIM adoption has been much slower than anticipated. There are two main reasons, technical and managerial.

The technical reasons can be broadly classified into three categories: The need for well-defined transactional construction process models to eliminate data interoperability issues; the requirement that digital design data be computable; the need for well-developed practical strategies for the purposeful exchange and integration of meaningful information among the building information model components.

The management issues cluster around the implementation and use of BIM. Right now, there is no clear consensus on how to implement or use BIM. Unlike many other construction practices, there is no single BIM document providing instruction on its application and use. Furthermore, little progress has been made in establishing model BIM contract documents. Several software firms are cashing in on the "buzz" of BIM and have programs to address certain quantitative aspects of it, but they do not treat the process as a whole. There is a need to standardize the BIM process and to define guidelines for its implementation. Another contentious issue among the AEC industry stakeholders (i.e., owners, designers, and constructors) is who should develop and operate the building information models and how the developmental and operational costs should be distributed.

To optimize BIM performance, either companies or vendors will have to find a way to lessen the learning curve of BIM trainees. Software vendors have a larger hurdle of producing a quality product that customers will find reliable and manageable and that will meet the expectations set by the advertisements. Additionally, the industry will have to develop acceptable processes and policies that promote BIM use and govern today's issues of ownership and risk management.

Researchers and practitioners have to develop suitable solutions to overcome these challenges and other associated risks. As a number of researchers, practitioners, software vendors, and professional organizations are working hard to resolve these challenges, it is expected that the use of BIM will continue to increase in the AEC industry.

<u>In the past, facilities managers have been included in the building planning process in a very limited way, implementing maintenance strategies based on the as-built condition at the time the owner takes possession. In the future, BIM modeling may allow facilities managers to enter the picture at a much earlier stage, in which they can influence the design and construction.</u>[10] The visual nature of BIM allows all stakeholders to get important information, including tenants, service agents, and maintenance personnel, before the building is completed. Finding the right time to include these

people will undoubtedly be a challenge for owners.

New Words and Expressions

procurement　　*n.*　　采购；获得，取得
specification　　*n.*　　规格；说明书；详述
collaborate　　*v.*　　合作
paradigm　　*n.*　　范例
rigorously　　*adj.*　　严厉地，残酷地
downstream　　*adj.*　　下游的，顺流而下的
formative　　*adj.*　　形成的
subcontractor　　*n.*　　转包商，分包者
stakeholder　　*n.*　　利益相关者
scenario　　*n.*　　方案；设想
cluster　　*v.*　　聚集

Notes

1. BIM can be viewed as a virtual process that encompasses all aspects, disciplines, and systems of a facility within a single, virtual model, allowing all design team members (owners, architects, engineers, contractors, subcontractors, and suppliers) to collaborate more accurately and efficiently than using traditional processes.

本句难点解析：这句话看起来很长，其实句子主体很短，即 BIM can be viewed as a virtual process，that 后面的部分用来修饰 BIM 到底是怎么样一个 virtual process，allowing 后面则是 BIM 的好处。

本句大意如下：BIM 可以被视为一个虚拟过程，它包含一个虚拟模型中设施的所有方面、学科和系统，可实现所有设计团队成员（业主、建筑师、工程师、承包商、分包商和供应商）更紧密地协作，并且比使用传统工艺更为有效。

2. As the model is being created, team members are constantly refining and adjusting their portions according to project specifications and design changes to ensure the model is as accurate as possible before the project physically breaks ground.

本句难点解析：这句话中 as 表示随着 create model 这个过程的进行，team members 是整句的主语，to 后面表示目的。break ground 中的 break 是一个多义词，表示"打破、突变、破裂"等意思，这里可以认为是"突变"的引申意义，表示工程项目从没有变成有的突变，因此可以解释为开工。

本句大意如下：随着模型的建立，团队成员不断地根据项目规范和设计变化对其份额进行细化和调整，以确保模型在项目实际开工之前尽可能准确。

3. BIM as a technology is still in its formative stage, and solutions in the market are continuing

to evolve as they respond to users' specific needs.

本句难点解析：本句表达的内容主要基于以下背景：如今，BIM 技术近年来被快速推广发展，但并未达到成熟应用的阶段，由于实际工程有各种各样细节的需求，因此市场上不断推出新的解决方案，使其在工程应用上能够更加方便。

本句大意如下：BIM 这种技术还处于形成阶段，为响应用户的特定需求，市场上的解决方案正在不断发展。

4. These models were created using detail-level information from subcontractors based on drawings from the designers.

本句难点解析：这句话表示修饰的部分非常多，层层相套。detail-level information 指的是"细节级别的信息"。需要注意的是，这些信息就是 BIM 技术和现有的建模方式区别很大的地方，正是因为引入了这些细节，才使得模型更加接近真实情况，能在施工进行之前对很多实际问题进行预测和优化。

本句大意如下：这些模型是根据设计师的图纸，使用分包商的细节级别的信息创建的。

5. Each critical clash was shared with the design team via the model viewer and a numbered collision log with a record of individual images of each collision per the architectural or structural discipline.

本句难点解析：这句话主体是 each critical clash was shared with design team

本句大意如下：每个重要的碰撞冲突都通过模型查看器和有编号的碰撞日志与设计团队共享。这个日志由建筑或结构学科的每个碰撞的单个图像记录。

6. Finally, the GC's commitment to updating the model to reflect as-built conditions provided the owner a digital 3D model of the building and its various systems to help aid operation and maintenance procedures down the road.

本句难点解析：这句话的主干是 the GC's commitment provided the owner a model to help aid operation and maintenance procedures。

本句大意如下：最后，总承包人的使命是更新反映竣工条件的模型，该使命为业主提供了建筑物的数字 3D 模型及其各种系统来帮助之后的辅助操作和维护过程。

7. The cost benefits to the owner were significant, and the unknown costs that were avoided through collaboration, visualization, understanding, and identification of conflicts early were in addition to the reported savings.

本句难点解析：本句表达的内容主要基于以下背景：BIM 技术由于它明显的经济优势，近年来被广泛地推广，深受业主的欢迎。

本句大意如下：业主的成本获益很明显，并且还有上文提及之外的，通过协作、可视化、理解、早期冲突识别这些途径避免的未知开销。

8. For example, equipment and material vendors offer designs associated with their products for the convenience of the lead designer in hopes of inducing the designer to specify the vendor's equipment.

本句难点解析：本句涉及一个关于许可的问题，虽然这种做法可能对企业有好处，但是如果设计不是由在项目所在地获得许可的设计者提供的，就可能出现许可的问题。

本句大意如下：例如，设备和材料供应商为了设计者的方便，提供了与其产品相关的设计，希望能引导设计者指定供应商的设备。

9. In cases where the data are incomplete or are submitted in a variety of scheduling and costing programs, a team member—usually a general contractor or construction manager-must re-enter and update a master scheduling and costing program.

本句难点解析：in cases 为 "如果" 的意思。

本句大意如下：如果数据不完整或在多种进度和成本计算程序中被提交，则团队成员（通常是总承包商或施工经理）必须重新输入和更新主调度和成本计算程序。

10. In the past, facilities managers have been included in the building planning process in a very limited way, implementing maintenance strategies based on the as-built condition at the time the owner takes possession. In the future, BIM modeling may allow facilities managers to enter the picture at a much earlier stage, in which they can influence the design and construction.

本句难点解析：句中的 picture 是一个非常形象的用法，通常 picture 指的是图片，文中 enter the picture 指的是进入影响工程项目。这两句话，对比了过去和将来设施管理人员对于项目影响的区别。

本句大意如下：在过去，设施管理人员在建筑规划过程中的影响非常有限，仅能基于业主获得使用权的竣工条件下实施维修策略。在未来，BIM 建模可能使设施管理人员在更早的阶段进入工程中，使其对设计和施工阶段产生影响。

Exercises

Translate the following phrases into Chinese.

1. building information model
2. project schedule
3. cost estimating
4. work put in place
5. overall risk
6. model viewer
7. investment analysis
8. scheduling data
9. industry stakeholder
10. AEC industry

Translate the following sentences into Chinese.

1. The architecture, engineering, and construction (AEC) industry has long sought techniques to decrease project cost, increase productivity and quality, and reduce project delivery time. Building information modeling (BIM) offers the potential to achieve these objectives.

2. The key benefit of a building information model is its accurate geometrical representation of

the parts of a building in an integrated data environment.

3. All the data reported in this section were collected from the Holder Construction Company (HCC), a midsize general contracting company based in Atlanta, Georgia (hereinafter referred to as the general contractor, or GC).

4. BIM risks can be divided into two broad categories: legal (or contractual) and technical. In the following paragraphs, key risks in each category are briefly discussed.

5. The productivity and economic benefits of BIM to the AEC industry are widely acknowledged and increasingly well understood. Further, the technology to implement BIM is readily available and rapidly maturing.

Part 4

Project Management and Case Study

Unit 16

Internationalization of Chinese Construction Enterprises

16.1 Abstract

Thirty-five international contractors from China were included by the Engineering News Record in the list of the Top 225 International Contractors in 2000. Although Chinese international contractors are increasingly playing a significant role in the global construction market, relatively few studies have been completed on their historical background and foray into the international arena. Apart from seeking to fill this lacuna, this paper also provides an analysis of these 35 Chinese international contractors to evaluate their achievements using the following performance indicators: International Revenue/Total Revenue, International Business Distribution, Overseas Management Structure, Involvement in Specialized Fields, and Overall Index of Internationalization. The analysis identified the top ten Chinese international contractors who are truly global in outlook. The study also suggests that the traditional multinational enterprise theories may not explain the development of Chinese international construction firms adequately.

16.2 Introduction

In recent years, China's construction enterprises were increasingly involved with international engineering projects, manpower services, and other cooperative projects overseas. According to the annual survey conducted by ENR (Engineering News Record), more than 30 of China's construction enterprises were included within the top 225 international contractors based on their construction revenues generated outside China in 2000. China's international construction enterprises (CICEs) are emerging as one of the strongest contenders in the field after international construction enterprises from the United States, UK, Japan and several other European countries. Although CICEs are growing recently to be more involved with global businesses, the amount of literature or analysis of their international performance available for study in this area is very limited.

In order to analyze the top CICEs, the construction industry and its relevant role in China's economy will first need to be reviewed. In this paper, the historical background and characteristics of CICEs are first introduced, followed by an analysis of the internationalization trends of CICEs in the global market. A model for measuring the degree of internationalization of CICEs is also proposed in the paper.

16.3 Construction Industry in China

The construction industry is one of the oldest traditional industries which formed the backbone in China's economy. The construction industry has developed rapidly as it is well recognized in China that infrastructural and urban development formed the most essential part for economic development.

In terms of its size, China's construction industry is relatively huge. As shown in Fig. 16-1, the annual production from the construction industry between 1980 and 1999 ranged from 4.3% (1980) to 6.6% (1999) of the Gross Domestic Product (GDP) with output value of up to Renminbi (RMB) 544.27 billion.

While the local construction industry played one of the most important pillars in China's domestic economy, many Chinese enterprises have also been involved in the international construction market. At the end of 2001, the cumulative dollar amount of overseas contracts since 1976 was reported to be U.S. $127.867 billion, of which the 2001 figure alone was U.S. $16.45 billion. Most of these overseas contracts were for civil engineering works in the developing countries.

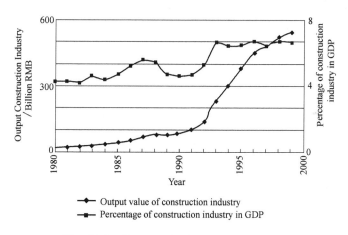

Fig. 16-1 Construction industry in China.

In China, enterprises in the construction industry are organized into three categories: State Owned Enterprises (SOEs); Urban and Rural Collectives (URCs); and Rural Construction Teams (RCTs). In 1999, there were more than 84250 construction enterprises in China employing over 23.65 million workers. These were made up of about 9394 SOEs with 6.9 million employees, 25442 URCs with 9.35 million employees and 49414 RCTs with nearly 7.4 million employees. The rapid growth in construction during the past 24 years has expanded the construction labor force, which increased from 9.8 million in 1980 to over 23.65 million in 1999. In recent years, there is an increasing trend for the emergence of private construction companies, due to the privatization of some URCs and RCTs. However, the number of private construction companies is still small relative to the entire construction industry.

Generally, the following characteristics of China's construction industry and enterprises may be identified:

16.3.1 Large Domestic Market and Huge Construction Work Forces

As shown in Fig. 16-2, the construction industry is closely related to the national fixed capital investment, which has been increasing rapidly along with China's economic growth. China's fixed capital investment in 1985 was RMB 254.3 billion, 65.1% of which were in construction and installation projects. In 1999, the total fixed capital investment had reached RMB 2975.46 billion, with 63.17% or RMB 1879.6 billion in construction and installation projects. In the foreseeable future, this trend will continue to remain high. On the other hand, China's construction work force continues to remain as the largest labor force in the working population. As mentioned above, almost 23.65 million people are working in the industry in 1999. This comprises of 6.9 million in SOEs, 9.35 million in URCs and 7.4 million in RCTs. Although China's construction industry contributes to employment opportunities to a large extent, it is still a very labor intensive industry that is not likely to change drastically in the near future. Fig. 16-3 shows that in 1999, the number of construction enterprises and employment increased to 1.5 times and 2.5 times that of 1980, respectively. Meanwhile, the total production was raised almost 36 times during the same period as shown in Fig. 16-4. The productivity in terms of production per employee increased 15 times as shown in Fig. 16-4. These trends suggest that improving the labor productivity of construction can remain a tough task to tackle in China.

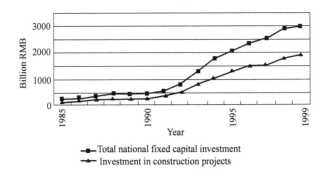

Fig. 16-2 National fixed capital and investments in construction/installations.

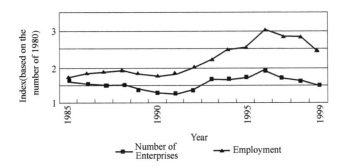

Fig. 16-3 Changes in China's construction enterprises and employment (1980 Index = 1).

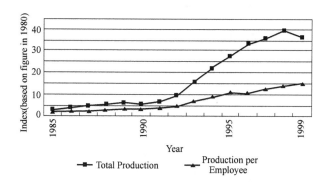

Fig. 16-4 Changes in total production and production per employee (1980 Index = 1).

16.3.2 Labor Intensive and Less Open Industry

China's labor-intensive construction industry, to some extent, does not presently rely on technological innovation. It is not as open an industry compared with other industries. This situation occurs because of its potential impact on employment, as well as the possible influence of reforms on State Owned and Urban Collective enterprises. Hence, foreign construction companies have limited access to the industry. Consequently, it is only five years after China's entry to the World Trade Organization (WTO) that foreign companies will be allowed to set up wholly owned enterprises in China.[1] However, although the construction industry is not as open as other industries, reforms are underway in many aspects.

Joint ventures and subcontracting are currently common within the various enterprises. There are very few private contractors in China. Although SOEs handled most of the construction in the past, their relative share is now decreasing. SOEs that comprise both the local units authorized by municipal governments and central ministry-affiliated enterprises have undertaken most of the construction of the infrastructure projects. For some years, notable progress has been achieved in reforming these enterprises in terms of commercial behavior, operational autonomy, and competitive bidding. However, SOEs still face many unresolved problems, i.e., poor management, use of old technology and an excessive labor force. The Urban Collectives and Rural Teams, on the other hand, have been developing fast.[2] The URCs and RCTs in 1999 accounted for over 71% of the construction labor force and produced about 62% of total construction output.

The rapid growth of the URCs and RCTs in China's construction industry helped contributed to the country's economic reforms towards a market-oriented system. While many SOEs are still in the process of reforms, the URCs and RCTs, who were set up in a market environment, have readily captured some market shares because of their low production costs, flexible labor force, profit-driven objective, and ease of movement from one city to another. However, the quality of the URCs and RCTs is relatively poorer because of their lower level of professional and technological management. In addition, the involvement of more RCTs in China's construction market, especially in the larger cities, has resulted in some social problems because of poor management and the large number of workers which the RCTs brought along with them.

16.3.3 Specialized Enterprises

Traditionally, the entire construction industry in China is dispersed in many economic fields, each of which is administrated by relevant government sectors. Generally, the enterprises may be categorized into construction of building and the construction of civil engineering projects. The former includes the construction of houses, office buildings, hospitals, and other buildings. The latter includes the construction of roads, highways, bridges, hydropower stations, thermo-power stations, nuclear power stations, irrigation works, and other infrastructures. Different types of projects are administrated by different government departments. Hence, each of the construction enterprises traditionally possesses specialty in a certain field. But along with the market-driven economy that is growing rapidly in China, construction enterprises have also reformed and diversified to include as many types of projects as possible.[3]

16.3.4 Delineation Between Design and Construction

China has a very well established system of design institutes. In 1999, there were nearly 12572 design institutes that employed 786370 employees, of whom 612027 were engineers or designers and the rest were supporting staffs. About 44% of employment in this field is administered by line ministries, with the rest managed by provincial and municipal governments. The need to develop an adequate construction supervision capacity was felt once contracts started to be awarded based on competitive bidding other than on an assignment basis as practiced earlier. Traditionally, there is no independent supervisory organization in China's construction industry. From the 1990s, some supervisory-based companies were set up gradually. Many of these companies were offshoots of state-owned design institutes, especially the larger ones. There are presently 277 A-class supervisory-based companies in China which were registered and approved by the former Ministry of Construction of China. In 2001, the registered supervision engineers in the former Ministry of Construction of China numbered 11330. Engineering consulting is a new but fast growing field in China. While design institutes still undertake some consulting work, they are not named as consultants in the market place as such.

16.3.5 Separation of Research and Development

Chinese construction enterprises usually do not have research and development (R&D) departments. A few construction R&D institutes are administrated by the line ministries, while the remaining are managed by provincial or municipal governments. Construction R&D works received relatively low emphasis in China compared with those in Japan, the UK, and the United States. In 2000, R&D expenditure in the construction industry was only RMB 530 million nation-wide, accounting for only 0.6% of the whole country's R&D expenditure. Apart from construction research institutes, most of the design institutes have their own research sections providing some supporting works, which construction enterprises do not have.[4]

16.4 Development of Chinese International Contractors

China's state owned enterprises dominate the domestic construction market as well as almost all the shares of China's enterprises in the international construction market. The development of Chinese international construction enterprises can be traced back to the 1950s when the Chinese government provided economic and technical aid to other developing countries. The historical penetration of the international construction market by Chinese enterprises can generally be divided into the following three stages.

16.4.1 Chinese Government's Economic and Technical Aid before 1979

The Chinese government's economic and technical aid provided prior to the 1970s refers to the financial donations to other developing countries to achieve the so-called objective of "liberation and independence of brother countries in the Third World". In this stage, the international involvement of Chinese construction firms is mainly for financial aid projects in some developing countries with funds provided by the Chinese government (see Table 16-1). These projects were agreed upon by the two governments and administered by the corresponding government authorities instead of independent enterprises. Essentially, these projects do not technically constitute part of the international construction market for the following reasons: (1) these are not motivated by the market place or profit-driven for the firms; (2) all project costs and other expenditure were funded by the Chinese government; (3) firms participated in the projects only and are not involved in any decision-making activities.

Table 16-1 Chinese government's economic and technical aid pre-1979

Period	Projects undertaken			Projects completed		
	Number	Investment over 10 million Renminbi	Investment over 100 million Renminbi	Number	Investment over 10 million Renminbi	Investment over 100 million Renminbi
1954–1963	234	32	1	101	6	—
1964–1970	555	70	10	313	31	3
1970–1978	509	101	8	470	59	7
Total	1307	202	19	884	96	10

Source: EOMC 1989.

However, during this stage, the Chinese construction enterprises involved gained basic information about the international market that helped to train many personnel when China opened her door to the world.[5]

16.4.2 Emergence of Chinese International Construction Enterprises

The Chinese construction industry started to reform in the early 1980s following China's open-door policy. On August 13, 1979, China's State Council introduced an Act which allows Chinese specialized companies to invest in other countries. In the construction industry, the government started to introduce regulations to help set the basic ground rules. At the enterprise level, the companies were gradually given the flexibility to operate as "commercial entities". Subsequently, several SOEs were separated from governmental departments, but they continued to work primarily for overseas financial aid projects until the mid-1980s. Soon after, SOEs at the central government level (under the direct administration of the corresponding Ministries) were able to obtain licenses issued by the former Ministry of Foreign Economic Relations and Trade to bid for projects in the international market.

The operations of these enterprises from then on were independent of financial aid from the Chinese government. They participated in international bidding, tendered for commercial projects and negotiated with their foreign counterparts. Their motivation soon turned to one that is profit-driven from going abroad.

In November 1978, China's first international construction enterprise—China Construction Engineering Corporation (formerly the China State Construction Engineering Corporation) was set up. Some of the largest SOEs were established soon after, such as: China Road and Bridge Corporation, China Civil Engineering Construction Corporation, China International Water and Electric Corporation, China National Complete Plant Import and Export Corporation, etc.

16.4.3 Development of Multinational Enterprises

Since the early 1990s, some of the largest state owned construction enterprises had gained experience in the international market. Subsequently, provincial-level and some other regional companies were allowed to obtain licenses for contracting overseas. The price-war among Chinese companies in some traditional markets in the developing countries (such as Pakistan, Iraq, and other Middle Eastern and African countries) also commenced. The more experienced and larger companies expanded their businesses to new countries rapidly. By 1994, several of the more established Chinese international construction companies had shaped up. Thereafter, the more profitable enterprises were encouraged to list in the stock market following a strict evaluation exercise, which means they would no longer be protected by the government. Between 1997 and 1998, many SOEs were completely separated from their respective government organizations. Large scale SOEs were supervised by the newly established Office of Large Scale State-Owned Enterprises under the State Council.

To be listed on the stock market appears to be one of the most important management and financial strategies for these enterprises. Hence, most construction enterprises are very keen on the stock market over the past few years, while the tough process of reforming the SOEs was still on going. As a strategy, construction enterprises off-loaded their unprofitable assets in favor of more favorable assets to form new share-holding companies after strict evaluation by the Committee of China Securities and Shares. By the end of 2001, however, few construction enterprises had successfully sold their shares through their share-holding companies.

Many Chinese international construction enterprises are presently struggling through the process of reforms to become multinational enterprises in the global market.

Where projects are concerned, the contracts successfully won by Chinese international companies overseas fall into the following categories:

1) Project funded through Chinese government loans or financial aid to developing countries.
2) Projects funded by loans from the World Bank or Asian Development Bank.
3) Projects obtained through government bilateral trade agreements.
4) Projects won through international bidding.
5) Projects obtained through local clients.
6) Projects obtained through local branch offices of Chinese enterprises.

In 2001, Chinese construction companies had expanded their businesses in more than 190 countries with 39400 new contracts with a total contract value of U. S. $16.45 billion (see Fig. 16-5).

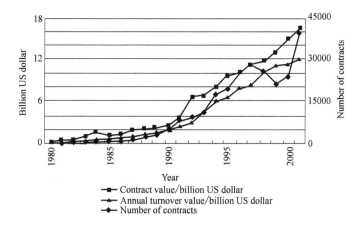

Fig. 16-5 Chinese construction enterprises in international market.

In 1999, the projects undertaken by Chinese international companies were distributed mainly in Asia and Africa, where about 73% of the total turnover value came from (See Table 16-2).

Overseas design and consultancy services were first offered by Chinese international construction companies in 1995. However, as shown in Fig. 16-6, the amount from design and consultancy services contracts was small relative to the overall contract value.

Although Chinese international construction enterprises have gained great stride in the global market, their scale of overseas operations is still small relative to their European, Japanese, and North American counterparts. According to the ranking of international contractors by Engineering News Record (ENR) in 2001, 35 Chinese enterprises were listed among the top 225 international contractors in the world (ENR 2001). The total international revenue of these 35 contractors was U.S. $5.383 billion in 2000.

While engaged with international contracts, equipment made in China was also exported in large volume. The total export value of equipment and material that accompanied overseas construction projects in 2000 was U. S. $875.59 million. These construction works contributed to economic development in the host countries through the completion of highways, dams, hydroelectric power stations, thermo power stations, transmission lines, and buildings.

The management mechanism commonly adopted by Chinese companies in overseas projects includes both the intercontracting and subcontracting arrangements. Interconstracting means that the project, upon the contract being signed between the contractor and client, will be contracted to the local branch office with a percentage for overhead charges.[6] The local branch office may contract the project wholly or partially to a project team who is from the same enterprise. Alternatively, the head office may directly contract the project to a project team with a percentage for overhead charges. Subcontracting, on the other hand, is to contract the project to other companies at a percentage for overhead charges. These companies may be Chinese, local, or from other countries. In both cases, the head office will maintain overall control of the project performance and provide the necessary assistance to the project team such as the working capital, major equipment allocation, etc.

Table 16-2 Annual turnover of Chinese international construction companies

(Unit: hundred thousand US dollar)

Region	1998				1999			
	Sum	Contracting projects	Labor service	Design and consultancy works	Sum	Contracting projects	Labor service	Design and consultancy works
Asia	6900	5322	1548	30.2	6247	4502	1711	34.1
Africa	2019	1871	144	4.3	2036	1828	203	5.5
Europe	489	239	246	4.2	306	126	171	9.2
Latin America	153	104	48	0.9	144	72	70	2.1
North America	322	111	204	6.8	331	104	225	1.9
Oceania and Pacific Islands	150	100	48	2.0	182	120	63	0.2
Others	52	36	16	—	59	44	14	0.1
Inside China	1688	1461	136	92.0	1930	1727	166	36.5
Total	11773	9243	2390	140.5	11235	8522	2623	89.6

Source: China Statistical Yearbook (2000).

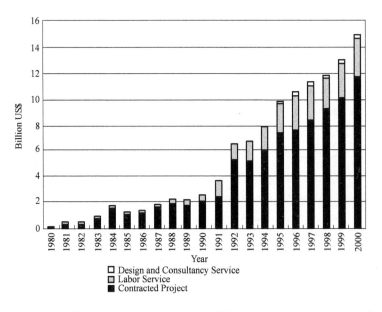

Fig. 16-6 Components of overseas contracts won by Chinese international construction companies.

16.5 Engineering News Record Top 35 Chinese International Contractors

ENR has annually ranked the Top 225 International Contractors according to construction revenues generated outside of each company's home country. In the 2000 ranking, 35 Chinese international contractors (including those from Hong Kong) were ranked among the top 225 international contractors in the world. In terms of the number of firms from a single country, China was ranked second after the U.S. (73 firms). Japan was ranked third with 21 firms, Germany fourth with 11 firms, and Italy fifth with 10 firms. The total share of Chinese firms, at U.S. $5.383 billion, accounted for 4.6% of the total billings (Table 16-3).

Although relatively more firms from China were ranked among the top 225 international contractors, their share of the total international billings is less than that of the other developed countries. Nevertheless, since all the Chinese firms were involved in the international construction market only over the past thirty years, the necessity of analyzing the performance of these firms is even more critical.

Table 16-3 Nationalities of top international contractor

Country	Number of firms		International billings		
	Rank	Number	Rank	U.S. dollars million	%
United States	1	73	1	24962.80	21.5
China	2	35	6	5383.80	4.6
Japan	3	21	5	8801.60	7.6
Germany	4	11	2	18162.60	15.7
Italy	5	10	10	3437.20	3.0
Spain	6	8	8	4405.20	3.8
France	7	7	3	15991.60	13.8
UK	8	7	4	9182.80	7.9
Korea	9	7	9	3611.80	3.1
Turkey	10	7	11	265.8	0.2
Canada	11	5	12	194.6	0.2
Netherlands	12	2	7	4522.40	3.9
All other countries	—	32	—	16985.30	14.7
All firms	—	225	—	115907.50	100

Source: ENR 2001.

16.6 Literature Review

Despite intensive research on internationalization and globalization of enterprises over the past decades, the determination of the degree of internationalization of a firm appears to remain an evolving one. ENR ranks international contractors according to the absolute amount of international revenue, which is one of the most acceptable ways to measure the degree of international performance of a firm. As suggested by Buckley, Dunning, and Pearce (1977), Stopford and Dunning (1983), and Daniels and Bracker (1989), a company's foreign sales or revenues are meaningful first-order indicators of its involvement in international business. The level of international revenues would form the basis to estimate the degree of internationalization of a firm. However, it could not reveal the overall degree of internationalization and various other related aspects for a firm.

The traditional multinational enterprise (MNE) theory, as established by Dunning (1977, 2000) along with other MNE theorists, suggests that the eclectic paradigm is a useful model for analyzing international production. The eclectic paradigm provides a framework to analyze the internationalization of a firm by taking into account its ownership (O), locational (L), and internalization (I) advantages. The transnationality index, adopted by the United Nations to evaluate multinational corporations, takes the average of three ratios: foreign assets to total assets, foreign sales to total sales and foreign employment to total employment as a measurement. Other literature suggests that the factors influencing the degree of internationalization of a firm may fall into three categories: performance, structural and attitudinal. Sullivan (1994) adopted five variables to measure the degree of internationalization of a firm: foreign sales as a percentage of total sales, foreign assets as a percentage of total assets, overseas subsidiaries as a percentage of total subsidiaries, psychic dispersion of international operations, and top managers' international experience. Another model which measures the degree of internationalization of a firm was suggested by Tong (2000) to include six quantifiable factors: pattern of international business management, financial management, marketing, human resource management, management structure, and the transnationality index adopted by the United Nations (2001).

By considering the advantages of internationalization of a firm in one direction and the impact in international operations in another direction, and taking into account the most significant aspects affecting the internationalization of a construction firm, an internationalization matrix may be constructed (Table 16-4).

Table 16-4 Internationalization matrix of firm

Factors	Ownership-Advantages	Locational-Advantages	Internalization-Advantages
Performance	Foreign/total revenue Foreign/total sales Foreign/total profit Technical specialty	Foreign employment Export/Import business	Equipment/material transfer Financial management Marketing Human resource management
Structural	Foreign/total assets	Overseas Subsidiaries Overseas branches	Pattern of business management Project management structure
Attitudinal	Top managers's international experience	Host country relationship	Psychic dispersion of international operation

As a rule of thumb, the analysis should be restricted to data that is achievable in general, and the measure adopted should enable an estimate to be made objectively. Hence, based on the proposed internationalization matrix, the analysis can be carried out to cover the following aspects:

1. The firm's international performance: Ownership or O-advantages

Because most of the Chinese construction enterprises are not listed publicly, information on their assets is not available. However, to estimate the firm's international performance, as a major part of its ownership advantages, its international revenue as a percentage of total revenue (O-advantage: IRTR) as listed by ENR, is adopted.

Unit 16 Internationalization of Chinese Construction Enterprises

2. The firm's international business distribution: Locational or L-advantages

To analyze the business distribution of a firm throughout the world, the countries where the firm is operating in are identified to reveal the locational advantages of the firm. However, Chinese firms may possess different locational advantages in terms of performance and structural factors (foreign employment, overseas subsidiaries, or branches, etc.) in different countries. Hence, the locational advantages of a particular country are adjusted by a coefficient. To reflect the different L-advantages of Chinese firms in different countries, this coefficient utilizes the total turnover of all Chinese international construction enterprises in a particular country as a percentage of the total turnover of all these enterprises throughout the world.[7] Consequently, a country where Chinese enterprises operate at a lower degree will receive a lower coefficient than the country where they are more involved in. Hence, the international business distribution (L-advantages: IBD) of a firm is estimated as the sum of the coefficients of those countries where the firm is working in.

3. The firm's overseas management structure: Internalization or I-advantages

Due to data privacy, it is difficult to measure all the internal factors of a firm. However, the pattern of business management and overseas management structure of a firm could demonstrate its internalization advantages to some extent.[8] Most of the overseas management pattern of Chinese international construction enterprises falls into the following categories:

1) Through local agent.
2) Representative office or liaison office.
3) Subsidiaries.
4) Joint-venture company.
5) Branch company (solely owned).

A firm with a representative office or liaison office operating overseas is given a one-point score, while each subsidiary or solely owned branch company overseas receives a two-point score. The overseas management structure (I-advantages: OMS) of a firm may be measured as the sum of these scores.

4. The firm's market involvement among different specialized fields: specialty-advantages

The international involvement of a construction firm is, to a great extent, restricted by its technical specialty advantages. The more diversified technical specialties a firm possesses, the more business shares it may obtain. ENR classifies the construction industry into ten specialized fields: general building, manufacturing, power, water supply, sewerage/solid waste, industrial process, petroleum, transportation, hazardous waste, and telecommunications. Thus, the number of such specialized fields that a firm is involved with out of the ten categories, can provide an estimate of its involvement in different specialized fields (S-advantages: ISF).

To estimate different aspects of internationalization of a firm on the same scale, three indices described above, i.e., L-advantages: IBD, I-advantages: OMS, and S-advantages: ISF, should be adjusted into a "0-1" scale, taking the maximum of a particular index of all firms as "1." A firm's overall index of internationalization (OII) sums O-IRTR, L-IBD, I-OMS, and S-ISF. The OII indicates the degree of internationalization of a firm on a relative basis.

16.7 Analysis of Engineering News Record Top 35 Chinese International Contractors

Following the methodology derived from the literature review described above, the analysis should combine the absolute performance of the firm in terms of its international revenue and its relative degree in terms of the four internationalization indices and the overall index of internationalization (OII).[9] The absolute international revenue of a firm forms the basis of an internationalization analysis, which reveals the firm's international presence. However, this ignores the other major aspects of the firm, such as the distribution of such revenue, its overseas management structure, as well as other factors. Hence, the international strength of a firm, which is derived from multiplying the absolute international revenue of the firm by its OII, may overcome this weakness (Table 16-5). The international strength of a firm reflects both its absolute level of international revenue and various other factors influencing its internationalization.

Table 16-5 shows the OII and the international strength of each firm among ENR's top 35 Chinese international contractors, along with their estimates of O-IRTR, L-IBD, I-OMS, and S-ISF.

The following analysis is drawn in two directions: horizontal analysis (i.e., comparison between firms) and vertical analysis (i.e., comparison of common attributes across firms).

16.7.1 Horizontal Analysis

A comparison of these 35 firms suggests that their business strategies appear to rely on different dominant markets.

1) The dominant business line lies in overseas markets. Due to historical reasons, some Chinese international firms developed their businesses mainly in the overseas market. This appears to contradict traditional MNE theories which suggest that enterprises could expand their businesses beyond the border only if they had already achieved a certain capacity in their home country. A few Chinese construction firms were mainly engaged in overseas markets. These include:

① China Civil Engineering Construction Corporation, Beijing, China.

② China National Complete Plant Import and Export Corporation, Beijing, China.

③ China Jiangsu International Economic and Technical Cooperation Corporation, Nanjing, China.

④ China International Water and Electric Corporation (CWE), Beijing, China.

⑤ Dongfang Electric Corporation, Chengdu, Sichuan Province, China.

⑥ China National Overseas Engineering Corporation, Beijing, China.

⑦ China Shanghai SFECO, Shanghai, China.

These firms only have a small portion or even no revenue from their domestic home markets in China. While this situation may change after China's entry into the WTO recently, it may remain the same for yet some time to come.

Unit 16 Internationalization of Chinese Construction Enterprises

Table 16-5 Overall internationalization index and international strengths of the top 35 Chinese international contractors

Number	Firm	Engineering News records ranking for 2000			International revenue to total revenue (IRTR) O-IRTR	Score for International business distribution (IBD) L-IBD	Score for overess mangement strikture (OMS) I-OMS	Score for trvolvement of specialized fields (ISF) S-ISF	Overall internationalization index (OII)		International strength of firm	Rank
		Engineering News records rank	Intenational reveme	Total revenue					On	Rank		
1	China State Const Engineering Corp	19	1278.7	4703.8	0.27	0.72	0.82	1.00	2.81	3	3598	1
2	China Hirbour Engineering Co.Grop	42	631.2	1862.4	0.34	0.93	1.00	0.71	2.99	2	1885	2
3	Paul Y-ITC Construction Holdings Ltd, HK	44	615.0	1457.0	0.42	0.58	0.79	0.57	2.36	11	1450	1
4	China Civil Engineering Construction Corp	70	272.2	288.0	0.95	0.48	0.71	0.29	2.42	10	659	4
5	China National Chenucal Engineering Corp	76	212.1	600.0	0.35	0.76	0.36	1.00	2.47	8	524	5
6	China Road and Bridge Corp	74	243.5	1181.1	0.21	0.57	0.57	0.43	1.77	17	432	6
7	China Jungst Int Ecoa-Tech Coop Corp	94	133.3	165.0	0.80	1.00	0.64	0.29	2.73	4	364	7
8	China Int Water and Electric Corp (CWE)	101	120.3	175.7	0.68	0.65	0.82	0.57	2.73	5	328	8
9	China Nat Complete Plant Inp. and Exp. Corp	87	166.2	166.2	1.00	0.17	0.46	0.29	1.92	15	319	9
10	China Metallurgkal Const. (Group) Corp	103	116.1	1550.3	0.07	0.54	0.86	1.00	2.47	9	287	10
11	China Shanghai SFECO	119	89.8	95.3	0.94	0.91	0.36	0.86	3.07	1	275	11
12	China National Oversess Engineering Corp	115	98.1	98.1	1.00	0.34	0.57	0.43	2.34	12	230	12
13	China Winbao Engineering Corp	125	78.5	87.9	0.89	0.67	0.64	0.43	2.64	6	207	13
14	CMEC	110	104.0	155.7	0.67	0.24	0.39	0.57	1.87	16	194	14
15	Shanghai Construction General Co	78	201.4	2224.0	0.09	0.51	0.18	0.14	0.92	31	185	15
16	China Railway Engineering Corp	88	165.9	3715.8	0.04	0.55	0.21	0.29	1.09	28	182	16
17	Dongfang Electric Corp	109	104.2	116.9	0.89	0.15	0.36	0.29	1.69	18	176	17
18	China Petroleum Engineering Constn Corp	114	100.0	167.3	0.60	0.21	0.29	0.29	1.38	22	138	18
19	China Elec Power Tech Inport and Export Corp	146	45.3	256.1	0.18	0.61	0.50	0.86	2.14	13	97	19
20	China Zhoogyun Engineering Corp	142	54.7	56.3	0.97	0.24	0.18	0.29	1.67	19	92	20
21	Harten Power Engineering Co.Ltd	123	82.1	111.3	0.74	0.09	0.11	0.14	1.07	29	88	21
22	China Wu Yi Corp	135	63.6	101.0	0.63	0.33	0.14	0.14	1.25	25	79	22
23	China Nat. Water Res. and Hydropower Eng.	136	61.7	1302.6	0.05	0.36	0.43	0.43	1.26	24	78	23
24	China Tianjin Int Eco and Tech Coop Corp	161	30.1	38.7	0.78	0.25	0.36	0.57	1.95	14	59	24
25	China Shenyang Int Eco and Tech Coop Corp	178	19.9	23.1	0.86	0.82	0.36	0.57	2.61	7	52	25
26	China Hamqin Chemical Engineering Corp	159	33.3	40.5	0.82	0.25	0.25	0.14	1.47	21	49	26
27	Zhejang Constuction Eng. Corp	150	41.1	461.3	0.09	0.53	0.36	0.14	1.12	27	46	27
28	China Rmlway Constrction Corp	154	37.9	3837.1	0.01	0.54	0.32	0.29	1.16	26	44	28
29	China Liaoning Int Corp Holdings Ltd.	173	24.4	31.3	0.78	0.09	0.36	0.43	1.65	20	40	29
30	China Dalian Int Cooperà Holdings Ltd	158	34.2	82.7	0.41	0.03	0.11	0.43	0.98	30	33	30
31	Sinopec Engineering Inc	132	65.2	585.2	0.11	0.18	0.07	0.14	0.50	34	33	31
32	Beijing Chang Cbeng Construkrioa Corp	193	12.4	651.4	0.02	0.56	0.45	0.29	1.33	23	16	32
33	China Huashi Enterpries corp	199	9.8	523.0	0.02	0.38	0.18	0.29	0.86	32	1	33
34	TEC China	196	10.9	40.1	0.27	0.01	0.04	0.29	0.61	33	7	34
35	Beijing Urban Construction Group Co Let	169	27.0	1474.5	0.02	0.03	0.04	0.14	0.23	35	6	35

Source: ENR 2001 (the interiction revemt and total reveine are in U.S. dollars million).

241

2) The business mainly relies on the domestic market. A majority of these firms mainly rely on the home market for businesses. They may be engaged in a small way in the international market as a strategy to diversify some risks or to seek other long term developments. These firms include:

① Shanghai Construction General Corporation, Shanghai, China.

② China Railway Engineering Corporation, Beijing, China.

③ China Metallurgical Group Corporation, Beijing, China.

④ China National Water Resources and Hydropower Engineering Corporation, Beijing, China.

⑤ China Railway Construction Corporation, Beijing, China.

⑥ Beijing Urban Construction Group Co. Ltd., Beijing, China.

3) The business developed and is balanced in both markets. Some firms have developed their businesses which are balanced in both the international and domestic market. These are also some of the most prominent companies in China. These firms include:

① China State Construction Engineering Corporation, Beijing, China.

② China Harbour Engineering Co. Group, Beijing, China.

③ Paul Y. -ITC Construction Holdings Ltd., Kowloon, Hong Kong.

④ China Road and Bridge Corporation, Beijing, China.

⑤ China National Chemical Engineering Corporation, Beijing, China.

4) Firms with high internationalization index but relatively lower foreign revenues OII is a relative index which reflects the various aspects of a firm related to international business development. Hence, the index may not be consistent with the absolute level of a firm's performance in the international market. One reason is that some firms, with the backing of the government, ventured into other countries not to pursue profit, i.e., they are not profit driven businesses. This probably reflects the malady in some of China's SOEs. It would appear that this situation will improve as economic reforms in China take effect further.

16.7.2 Vertical Analysis

1. International business distribution

From an overview of the international business distribution of these 35 firms, it can be observed that some of them have concentrated their businesses in a few key countries, while others have developed their scope of business in many countries. This may be related to the firm's business strategy to either maintain a few important overseas markets, where they may have operated for a few years, or to expand their businesses in many countries to capture more potential opportunities.[10] Examples of these two groups of companies are shown below.

Businesses developed in a few key countries:

1) Harbin Power Engineering Co. Ltd., Harbin, China.

2) China Wu Yi Corporation, Fuzhou, China.

3) China Zhongyuan Engineering Corporation, Beijing, China.

4) China Dalian International Cooperation Holdings Ltd., Dalian, China.

5) Beijing Urban Construction Group Co. Ltd., Beijing, China.

Businesses developed in many countries:

1) China State Construction Engineering Corporation, Beijing, China.

2) China Harbour Engineering Co. Group, Beijing, China.

3) China National Chemical Engineering Corporation, Beijing, China.

4) China Jiangsu International Economic and Technical Cooperation Corporation, Nanjing, China.

5) China Shanghai SFECO, Shanghai, China.

2. Overseas management structure

Due to the nature of construction works, most of the Chinese construction firms operate their overseas businesses through representative offices on a project basis. Some of these firms may set up a local branch office or joint venture company to pursue interests in countries where restrictions are imposed. For example, in some countries, the local or joint venture companies may enjoy a 7% discount off the bidding price. Hence, this may force the foreign firm to set up a joint venture with a local firm. In addition, the foreign direct investments (FD) to other countries by Chinese international construction firms are not very significant. It is only in a few countries where they have operated for many years with an intent for a longer stay, may they then establish a subsidiary or solely owned company. For example, as one of its overseas business strategies, China State Construction Engineering Corporation usually develops and operates its overseas businesses on a project basis through its 19 representative offices throughout the world. For example, because of existing business opportunities in Singapore and the Southeast Asian market, the Singapore branch company is an active subsidiary of China State Construction Engineering Corporation.

3. Involvement of specialized fields in the construction industry

Most Chinese international construction firms have focus on general building projects overseas. A few firms, with their specialized background in China, have executed other specialized projects. For example, apart from general building projects, China Civil Engineering Construction Corporation (formerly under the administration of the Ministry of Railway, China) had engaged in transportation projects; China International Water and Electric Corporation (formerly under the administration of the Ministry of Water Resources, China) is adept in power and water supply projects; China Petroleum Engineering Construction is skilled in industry/petroleum projects, and so on.

However, a few Chinese international construction firms are developing their own specialty in a more diversified manner. For example, projects undertaken in 2000 by China State Construction Engineering Corporation covered seven specialized fields out of ten, while China Harbor Engineering Co. Group was involved in five fields; China Metallurgical Construction Corporation in seven specialized fields; China Shanghai SPECO in six fields, etc.

16.8 Conclusion

Combining both the absolute measure of international business performance and the relative OII of a firm, the truly global Chinese contractors may be identified. As shown in Table 16-6. The truly global Chinese contractors are ranked according to their international strengths. They exhibit higher OIIs and international revenues among the 35 firms listed in ENR's top 225 international contractors. The top ten Chinese contractors who may be regarded as truly global contractors both in absolute and relative terms include:

1. China State Construction Engineering Corporation.
2. China Harbour Engineering Co. Group.
3. Paul Y. -ITC Construction Holdings Ltd., HK.
4. China Civil Engineering Construction Corporation.
5. China National Chemical Engineering Corporation.
6. China Road and Bridge Corporation.
7. China Jiangsu International Economic-Technical Cooperation Corporation.
8. China International Water and Electric Corporation (CWE).
9. China National Complete Plant Import and Export Corporation.
10. China Metallurgical Construction (Group) Corporation.

Table 16-6 Truly Global Chinese Contractors

Rank	firm	Engineering news records rank	Overall internationalization index	Rank by overall internationlization index	International strength
1	China State Const Engineering Corp.	19	2.81	3	3598
2	China Harbour Engineering Co.Group	42	2.99	2	1885
3	Paul Y.—ITC Construction Holdings Ltd., HK	44	2.36	11	1450
4	China Civil Engineering Construction Corp.	70	2.42	10	659
5	China National Chemical Engineering Corp.	76	2.47	8	524
6	China Road and Bridge Corp.	74	1.77	17	432
7	China Jiangsu Int.Econ-Tech.Coop.Corp.	94	2.73	4	364
8	China Int. Water and Electric Corp.(CWE)	101	2.73	5	328
9	China Nat. Complete Plant Imp. and Exp.Corp.	87	1.92	15	319
10	China Metallurgical Const.(Group)Corp.	103	2.47	9	287

There are, however, a number of limitations in this study. First, due to the unavailability of data, the study only examined some aspects of a firm's internationalization exercise, instead of all aspects. Hence, the study is limited in its scope as defined in the paper. Second, the data used in this study was developed based on a few major professional/governmental publications, whose reliability needs to be verified through other sources. Finally, because of some special characteristics of China's international construction enterprises, especially those of the reformed SOEs in China, the traditional MNE's theories may not be able to explain the Chinese situation adequately. Although more factors may need to be incorporated in the study, further theoretical study and more comprehensive research effort are recommended. Following China's open door policy, further study is also recommended to examine the international marketing strategies adopted by Chinese construction enterprises in the global market when they gradually compete against their more established North American, Japanese, and European counterparts.

Unit 16　Internationalization of Chinese Construction Enterprises

 New Words and Expressions

  internationalization　　*n.*　　国际化
  backbone　　*n.*　　支柱
  literature　　*n.*　　文献；著作
  bidding　　*n.*　　投标；出价
  delineation　　*n.*　　描述
  expenditure　　*n.*　　支出，花费
  subsidiary　　*adj.*　　附属的；辅助的
  subcontract　　*n.*　　分包合同
  restriction　　*n.*　　限制；约束；束缚

 Notes

  1. Consequently, it is only five years after China's entry to the World Trade Organization (WTO) that foreign companies will be allowed to set up wholly owned enterprises in China.

  本句难点解析：it is…that 是强调句的句型，强调了 only five years。

  本句大意如下：因此，中国加入世贸组织（WTO）后仅五年，就允许外国公司在中国设立全资企业。

  2. The Urban Collectives and Rural Teams, on the other hand, have been developing fast.

  本句难点解析：on the other hand 为插入语，也可以放在句首，但是在英文的表述中，类似的短语常常会放在句中。

  本句大意如下：另一方面，城市集体建筑企业和农村集体建筑企业发展很快。

  3. But along with the market-driven economy that is growing rapidly in China, construction enterprises have also reformed and diversified to include as many types of projects as possible.

  本句难点解析：market-driven economy 解释为"市场经济驱动"。

  本句大意如下：但随着中国的市场经济驱动，建筑企业也进行了改革和多元化发展，将尽可能多的项目纳入其中。

  4. Apart from construction research institutes, most of the design institutes have their own research sections providing some supporting works, which construction enterprises do not have.

  本句难点解析：本句的主体为 design institutes have their own research sections，providing 后面表示目的，which 后面修饰 supporting works。

  本句大意如下：除了建筑研究机构，大多数设计院都有自己的研究部门，这些机构和部门可以提供一些施工单位没有的支撑工作。

  5. However, during this stage, the Chinese construction enterprises involved gained basic information about the international market that helped to train many personnel when China opened her door to the world.

  本句难点解析：句中 involved 修饰 construction enterprises，gained 是句子的动词。句子

的主体为 construction enterprises gained basic information to train many personnel。

本句大意如下：然而，在这个阶段，涉及的中国的建设企业获得了国际市场的基本信息，帮助中国在向世界敞开大门的过程中培养了许多人才。

6. Intercontracting means that the project, upon the contract being signed between the contractor and client, will be contracted to the local branch office with a percentage for overhead charges.

本句难点解析：句中 upon the contract being signed between the contractor and client 可认为是插入的解释，不是句子的主体部分。overhead charges 意思是企业一般管理费。整句句子为了解释什么是 intercontracting（分包）。

本句大意如下：分包意味着项目在承包商和客户之间签订合同的基础上，将与当地分支机构签订合同，并支付一定比例的间接管理费。

7. To reflect the different L-advantages of Chinese firms in different countries, this coefficient utilizes the total turnover of all Chinese international construction enterprises in a particular country as a percentage of the total turnover of all these enterprises throughout the world.

本句难点解析：本句的后半句比较长，其实结构并不复杂，第一个 of 修饰第一个 total turnover，最后一个 of 修饰第二个 total turnover，句中提到的系数就是这两个 total turnover 的比值。

本句大意如下：为了反映中国企业在不同国家的不同地域优势，这一系数利用了在某一个国家中，所有中国国际施工单位的总营业额，占全球所有这些施工单位总营业额的百分比。

8. However, the pattern of business management and overseas management structure of a firm could demonstrate its internalization advantages to some extent.

本句难点解析：to some extent 译为"在一定程度上"。

本句大意如下：但企业的管理模式和海外管理结构在一定程度上可以体现其内在的优势。

9. Following the methodology derived from the literature review described above, the analysis should combine the absolute performance of the firm in terms of its international revenue and its relative degree in terms of the four internationalization indices and the overall index of internationalization (OII).

本句难点解析：combine A and B, 把 A 和 B 相结合。需要结合的部分一个是绝对的表现，由其后 in terms of 后的短语修饰，另一个部分是相对的程度，由第二个 in terms of 后的短语修饰。

本句大意如下：按照上述文献综述的方法，分析应结合企业在国际收入方面的绝对表现及其在四个国际化指数和总体国际化指数（OII）间的相对程度。

10. This may be related to the firm's business strategy to either maintain a few important overseas markets, where they may have operated for a few years, or to expand their businesses in many countries to capture more potential opportunities.

本句难点解析：本句介绍了两种业务战略，由 either…or…连接。

本句大意如下：这可能与公司的业务战略有关，维持几个已经运营了一些年的重要海外市场，或者在许多国家扩大业务，以吸引更多的潜在机会。

Exercises

Translate the following phrases in to Chinese.

1. international contractor
2. China's international construction enterprises (CICEs)
3. cumulative dollar
4. corresponding government authorities
5. transnationality index
6. the degree of internationalization of a firm
7. overseas management structure
8. China Civil Engineering Construction Corporation
9. comprehensive research effort
10. compete against counterparts

Translate the following sentences into Chinese.

1. Thirty-five international contractors from China were included by the Engineering News Record in the list of the Top 225 International Contractors in 2000.

2. The construction industry has developed rapidly as it is well recognized in China that infrastructural and urban development formed the most essential part for economic development.

3. China was ranked second after the U. S. (73 firms). Japan was ranked third with 21 firms, Germany fourth with 11 firms, and Italy fifth with 10 firms.

4. The traditional multinational enterprise (MNE) theory, as established by Dunning (1977, 2000) along with other MNE theorists, suggests that the eclectic paradigm is a useful model for analyzing international production.

5. Combining both the absolute measure of international business performance and the relative OII of a firm, the truly global Chinese contractors may be identified.

Unit 17
Project Management and Administration

■ 17.1 The Need for Project Management

The essential focus of a construction company is its field projects. These are, after all, the heart and soul of any such business enterprise. If a project is to meet its established time schedule, cost budget, and quality requirements, close management control of field operations is a necessity.[1] Project condition such as technical complexity, importance of timely completion, resource limitations, and substantial costs put great emphasis on the planning, scheduling, and control of construction operations. Unfortunately, the construction process, once it is set into motion, is not a self-regulating mechanism and requires expert guidance if events are conformed to plans.

It must be remembered that projects are one-time and largely unique efforts of limited time duration which involve work of a non-standardized and variable nature. Field construction work can be profoundly affected by events that are difficult, if not possible, to anticipate. Under such uncertain and shifting conditions, field construction costs and time requirements are constantly changing and can seriously deteriorate with little or no advance warning. Skilled and unremitting management effort is not only desirable, but is absolutely imperative for a satisfactory result.

■ 17.2 Project Organization

All construction projects require some field organization, although large jobs will obviously require considerably larger organizations than smaller jobs. Terminology differs somewhat from one construction firm to another and organizational patterns vary, but the following description is more or less representative of current trade practice.

The management of field construction is customarily run on a project basis, with a project manager being made responsible for all aspects of the work. Project management cuts across functional lines of the parent organization, and the central office acts in a service role to the field projects. Working relations with a variety of outside organizations, including architect-engineers, owners or owner representatives, subcontractors, material and equipment dealers, regulatory agencies, and possibly labor unions, are an important part of guiding a job through to its conclusion. Project management is directed toward pulling together all the diverse elements involved into a going venture with the common objective of project completion.

The form and extent of a project's organization depend on the nature of the work, size of the project, and type of construction contract. A firm whose jobs are not particularly extensive will have essentially all office functions, such as accounting, payroll, and purchasing, concentrated in its main or area office. Only larger projects can justify the additional over-head necessary to carry out the required office tasks in a field office on the jobsite. Extensive projects frequently support a substantial field management team, the extent of this depending on the nature of the work, its geographical location, and the type of contract. For example, a large cost-plus contract might well have all associated office functions performed at the project site. A project management staff is customarily developed along much the same lines as contractor's main operating organization.

■ 17.3 The Project Manager

The project manager organizes, plans, schedules, and controls the fieldwork and is responsible for getting the project completed within the time and cost limitations. He acts as the focal point for all facets of the project and brings together the efforts of those organizations having inputs into the construction process. He coordinates matters relevant to the project and expedites project operations by dealing directly with the individuals and organizations involved. In any such situation where events progress rapidly and decisions must be consistent and informed, the specific leadership of one person is needed. Because he has the overall responsibility, the project manager must have broad authority over all elements of the project. The nature of construction is such that he must often take action quickly on his own initiative, and it is necessary that he be empowered to do so. To be effective the project manager must have full control of the job and be the one voice that speaks for the project. Project manager is a function of executive leadership and provides the cohesive force that binds together the several diverse elements into a team effort for project completion.[2]

When small contracts are involved a single individual may act as project manager for several jobs simultaneously. Fig. 17-1 illustrates such an organization plan. Larger projects normally will have a full-time project manager who reports to a senior executive of the company. The manager may have a project team to assist him or he may be supported by a central office functional group. Such an arrangement is shown by Fig. 17-2.

The project manager must have expertise and experience in application of specialized management techniques for the planning, scheduling, and cost control of construction operations. These are procedures that have been developed specifically for application to construction projects. Because much of the project management system is usually computer based, the project manager must have access to adequate computer support services.

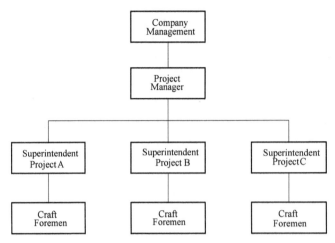

Fig. 17-1　Project organization, small projects.

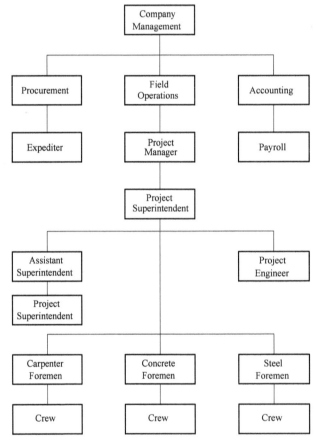

Fig. 17-2　Project organization, large project.

■ 17.4　The Project Superintendent

Project management and field supervision are quite different responsibilities. The day-to-day direction of project operations is handled by a site supervisor or project superintendent. His duties involve supervising and directing the trades, coordinating the subcontractors, working closely with the owner's or architect-engineer's field representative, checking daily production, and keeping the work progressing smoothly and on schedule. He is responsible for material receiving and storage, equipment scheduling and maintenance, project safety, and job records and reports.

Centralized authority is necessary for the proper conduct of a construction project, and the project manager is the central figure in that respect. Nevertheless, some freedom of action by the field superintendent is required in field construction work. In practice, construction project authority is wielded much as a partnership effort, with the project manager and project superintendent functioning much as allied equals.[3] Notwithstanding this, however, top company management must make a clear delegation of authority to the project manager and must also make sure the duties and responsibilities of each are explicitly understood.

On large projects, the field staff normally includes a field engineer who reports either to the project manager or the project superintendent. The engineer is assigned such responsibilities as project scheduling, progress measurement and reporting, progress billing, shop drawings, keeping of job records and reports, cost studies, testing, job engineering and surveys, safety and first aid, and payrolls.

17.5 Jobsite Computers

There is substantial use of microcomputers now being made at the jobsite itself. These computers are used as word processors and to store, transmit, and generate job information pertaining to time schedules, labor and equipment costs, time status reports payroll data, estimating, materials, and subcontractors. These machines functions in a stand-alone capacity or can be integrally linked to the company's mainframe computer. Field personnel with only minimal training can use these machines effectively. There can be substantial advantage in having such support equipment located on larger jobs, principally because it affords project management personnel with the capacity to develop needed information and to collect project data promptly and easily. With the proper software, the machine can substantially increase the productivity and superintendent with accurate and current job information.

17.6 Aspects of Project Management

In general terms, project management might be described as the judicious allocation and efficient usage of resources to achieve timely completion within the established construction budget. The resources required are money, manpower, equipment, materials, and time. Considerable management effort is required if the contractor is to meet its construction objectives. The achievement of a favorable time-cost balance by the careful scheduling and coordination of labor, equipment, and subcontractors, and the maintaining of a material supply to sustain this schedule requires endeavor and skill. The project organization must blend the quantitative techniques of scientific management together with the subjective ingredients of experienced judgment and intuition into an effective and efficient operating procedure. Astute project management requires at least as much art as science—as much human relations as management techniques.

The details of a job management system depend greatly upon the contractual arrangements with the owner. Basic to any contractor's project management system, however, is the control of project time and cost during the construction period. Before field operations begin, a detailed time schedule of operations and a comprehensive construction budget are prepared. These constitute the accepted time and cost goals that will be used as a flight plan during the actual construction process. After the project has been started, monitoring systems are established that measure actual costs and progress of the work at periodic intervals. The reporting system provides progress information that is measured against the programmed targets. Comparison of field costs and progress with the established plan quickly detects exceptions that must receive prompt management attention. Data from the system can be used to make corrected forecasts of costs and time necessary to complete the work.

17.7 Field Productivity

Project management is vitally concerned with field productivity because this is the measure that determines whether or not the project will be completed within the established cost budget and time schedule. Field productivity has declined seriously during the past several years. How to motivate workers and get more work for the money is a vital aspect of project management.

The CICE study disclosed that an appreciable proportion of a construction worker's time is completely non-productive, and at least one-half of the time wasted is caused by poor job management. Some of the aspects of this are demotivated workers, lack of material management, poorly trained supervisors, failure to utilize modern time and cost management systems, poor communications, and lack of teamwork. To overcome these negative factors, there must be a commitment by company management to establish a detailed program of action.

Many things can demotivate a craftsman, but the CICE study reports that some of the most common complaints are: materials unavailable, unsafe working conditions, redoing work already completed, unavailability of tools or equipment, lack of communication, and disrespectful treatment by supervisors.[4] These are all factors that can be minimized or eliminated by proper management action. Associated with workers being "turned off" is absenteeism and job turnover, both factors in poor productivity.

It is a surprising fact that few construction companies make any attempt to give their foremen and superintendents any type of supervisory training. There are now many such training programs available through contractor associations, consultants, and technical schools. In this regard, the Associated General Contractors of America sponsors the Supervisory Training Program (STP) and the Associated Builders and Contractors have program called Training Orientation Program for Supervisors (TOPS). These programs include such topics as communications, work planning, motivation, leadership, cost-effectiveness, human relations, safety, contact documents, and problem solving.

The CICE study also disclosed that scheduled overtime can be counterproductive and inefficient. Scheduled overtime does not refer to occasional periods when more than the usual 40 hours per week are worked. Scheduled overtime refers to where excess hours are made a part of the usual workweek for extended periods. As such overtime increases, productivity actually decrease.

■ 17.8　Project Administration

Project administration refers to those actions that are required to achieve the established project goals. These involve duties that may be imposed by the construction contract or that are required by good construction and business practice. The efficient handling, control, and disposition of contractual and administrative matters in a timely fashion are of paramount importance for a smooth-running job.[5]

Specifically, project administration refers to those practices and procedures, usually routine, that keep the project progressing in the desired fashion. Whatever is required to provide the project, efficiently and in a timely manner, with the materials, labor, equipment, and services required lies within the general jurisdiction of project administration.

■ 17.9　Project Meetings

On larger projects, it is common practice for representatives of the owner and the architect-engineer to meet with project personnel of the general contractor and major subcontractors before field operations actually begin. The purpose of this meeting is to establish important ground rules and to call the contractors' attention to certain critical areas. It permits key personnel of both sides to be introduced to each other and allows the parties to arrive at clear understandings concerning a variety of mutual concerns such as construction site surveys, submittal of shop drawings, sampling and testing, project inspection, the authority of the owner's field representative, payment requests, handling of claims, change orders, jobsite security, owner-furnished equipment, temporary facilities, and similar matters.

The agenda of such meetings usually includes reminders to the contractor concerning insurance certificates, required permits, cost breakdowns, lists of subcontractors, construction schedule, schedule of owner payments, and other actions required before the start of construction. Such contractual provisions as the completion date, liquidated damages, bonus clauses, safety program, and extensions of time may also be discussed. This pre-construction meeting gives the contractor the opportunity to raise questions and clear up misunderstandings.

Once construction has started, regular jobsite meetings are standard practice on larger projects. Such meetings are run by the project manager with representatives of the owner, material vendors, and subcontractors attending. Minutes of these meetings are kept and copies distributed to interested parties. At such meetings job progress is discussed, with trouble spots being identified and corrective action planned. These regular meetings are very valuable in the sense that all parties concerned are kept fully informed concerning the current job status and impressed with the importance of meeting their own obligations and commitments.[6] Such face exchanges are invaluable in resolving misunderstandings and quickly getting to the real source of project problems and difficulties.

17.10 Schedule of Owner Payments

A common contract provision requires the contractor to provide the owner with an estimated schedule of monthly payments that will become due during the construction period. This information is needed by the owner so that cash will be available as needed to make the necessary periodic payments to the contractor. Because the owner must often sell bonds or other forms of securities to obtain funds with which to pay the contractor, it is important that the anticipated payment schedule be as accurate a forecast as the contractor can make it.

The most accurate basis for determining such payment information is the project time schedule that is established by the contractor either before field operations start or very early in the construction period. By establishing the total costs associated with each scheduled segment of the project and by making some reasonable assumption concerning how construction cost varies with time (a linear relationship is often assumed), a reasonably accurate prediction of the value of construction in place at the end of each month can be made.[7] By taking retainage into account, these data can be reduced to an estimated schedule of owner monthly payments. These values may have to be subsequently revised as the project progresses.

17.11 Shop Drawings

The working drawings and specifications prepared by the architect-engineer, although adequate for job pricing and general construction purposes, are not suitable for the fabrication and production of many required construction products. Manufacture of the necessary job materials and machinery often requires that the contract drawings be amplified by detailed shop drawings[①] that supplement, enlarge, and clarify, the contract design. Such descriptive technical submissions are prepared by the producers or fabricators of the materials and are submitted to the general contractor and thence to the architect-engineer for approval before the item is supplied. This procedure also applies to materials provided by subcontractors, in which case shop drawings are submitted to the general contractor through the subcontractor. Shop drawings are required for almost every product that is fabricated away from the building site. To illustrate, in building construction, shop drawings must be prepared

---

① The term "shop drawings" includes such fabrication, erection, and setting drawings; manufacturer's standard drawings or catalog cuts; performance and test data; wiring and control diagrams; schedules; samples; and descriptive data pertaining to material, machinery, and methods of construction as may be necessary to carry out the intent of the contract drawings and specifications.

for everything from reinforcing steel and metal door frames to millwork and finish hardware. The general contractor must sometimes prepare shop drawings covering work items or appurtenances it has designed and will fabricate.

■ 17.12　Approval of the Shop Drawings

When shop drawings are first received from a supplier, the contractor is responsible for checking them carefully against the contract drawings and specifications. The shop drawings are then forwarded to the architect-engineer for examination and approval. The checking and certifying of these drawings is properly the responsibility of the architect-engineer, since they are basically a further development and interpretation of the design, and the final verification is the designer's responsibility. Approval of shop drawings can possibly place legal responsibility upon the architect-engineer if error in the drawings is later found to be the proximate cause of damages. Possibly to minimize their potential liability, many architect-engineers no longer use the word "approved" when indicating that shop drawings have been found satisfactory. Rather, "reviewed" "no exceptions taken" "accepted" or "examined" are now sometimes used.

A sufficient number of copies of each drawing must be provided so that there are enough for distribution to all interested parties. After the approved drawings have been returned by the architect-engineer, the contractor notes the nature of any comments or corrections and routes copies as necessary, returning at least one copy to the supplier. Occasionally the shop drawings must be redone and resubmitted to the architect-engineer. Because the material supplier must receive a copy of the approved shop drawings before placing the order into production, it is important that the submittal-approval-return process be expedited as much as possible. Otherwise, materials may not be delivered to the project by the time they are needed. It is usual for the contractor to establish some form of check-off and reminder system with regard to shop drawings to guard against oversight or delay in the approval process.

It is important to note that only a qualified approval is given to shop drawings by architect-engineers. Such approval relates only to conformance with the design concept and overall compliance to the contract drawings and specifications. Checking does not include quantities, dimensions, fabrication methods, or construction techniques. Approval of shop drawings by the architect-engineer does not relieve the contractor of its responsibility for errors or inadequacies in the shop drawings or for any failure to perform the requirements and intent of the contract documents. Approval of shop drawings does not authorize any deviation from the contract unless the contractor gives specific notice of the variance and receives express permission to proceed accordingly. However, as long as there is no explicit disagreement between the shop drawings and the construction contract, approval is usually binding in the event of a subsequent dispute over design requirements. For obvious reasons, it is important that contractors carefully check their project shop drawings. They cannot act merely as go-betweens with suppliers and architect-engineers. An interesting aspect of approved shop drawings is that they are not usually considered to be contract documents, and they do not modify, or extent the obligation of either party to the construction contract.

17.13 Quality Control

Quality control during field construction is concerned with ensuring that the work is accomplished in accordance with the requirements specified in the contract. The architect-engineer establishes the criteria for construction, and the quality control program checks contractor compliance with those standards. A field quality control program involves inspection, testing, and documentation for the control of the quality of materials, workmanship, and methods. On a given project, the quality control program may be administered by the architect-engineer, owner, consultants, prime contractor, or construction manager. The inspector has no authority to give directions, render interpretations, or change the contract requirements. The inspector is not present to manage the job, direct the work, or to relieve the contractor from any of its obligations. He cannot tell the contractor what to do or how to do it, nor interfere in field operations unless it is to prevent something from being done improperly. The function of the inspector is to observe the construction process and to ensure compliance with contract requirements by the contractor.

The providing of field inspection services can significantly increase the architect-engineer's vulnerability to a wide range of potential liabilities. To illustrate, the responsibility for field inspection has led the liability of architect-engineers for construction defects caused by the contractor's failure to follow contract requirements.[8] Design professionals have been made responsible for contractor's construction methods and have been held liable to workers and members of the public injured by construction operations. As a result of this, there has been a trend for architect-engineers to withdraw as much as possible from any responsibility for site operations. Extensive revisions have been made to the contract language that defines their field inspection responsibilities. The word "supervision" has proven to be especially troublesome for architect-engineers because it suggests that the designer has some degree of control and thus responsibility for the contractor's day-to-day operations. By the terms of most of the usual design contracts between owner and architect-engineer, the designer accepts only very limited responsibility for field operations. If the architect-engineer does agree to administer the project quality control program, it will assign a full-time employee to the project to provide administrative and surveillance services. Laboratory and field testing may be done by this person, or a commercial testing laboratory may be engaged to perform the specialized work.

Many public agencies and corporate owners establish their own internal quality control programs to monitor their ongoing construction projects. Where there is a continuous construction program or where very large and complex projects are involved, the owner may establish a functional department within its overall organization that acquires trained personnel and develops standards for quality control application in the field.

Owners sometimes hire specialized consultants to provide field quality control services. In this context, a consultant is a technical firm such as a testing laboratory, specialized consulting engineer (not the project designer), or construction management firm that provides quality control

Unit 17 Project Management and Administration

services. Such consultants are frequently employed on extremely complex construction or on projects that have specialized and highly technical quality control requirements. These firms are entirely independent of the architect-engineer and contractor and are usually hired before the design is completed to provide advance quality assurance input and advice.

During recent years, some public and private organizations have required prime contractors to take a more active role in the control of project quality by making them develop and manage their own quality control programs. Construction contracts with these owners require the contractor to maintain a job surveillance system of its own and to perform such inspections that will assure the work performed conforms to contract requirements. The contractor is required to maintain and make available adequate records of such inspections. Owner representatives monitor the contractor's quality control plan and make spot-check inspections during the construction process. Under such contracts, the contractor is required to provide significant and specific inspection and documentation to satisfy both itself and the owner that the work being performed meets the contract requirements. In the usual case, the contractor is required to report on a daily basis the construction progress, problems encountered, and corrective action taken, and to certify that the completed work conforms to the drawings and specifications. The owner is responsible for final inspection and may inspect at any other time deemed necessary to ensure strict compliance with the contract provisions.

The last aspect of quality control is the final inspection, field acceptance testing, and start-up of the facility. The prime contractor, appropriate subcontractors, and manufacturer's representatives start up the project equipment and systems, with the owner checking all control and instrument operation. This includes simulation of operating and emergency conditions. The facility is turned over to the owner with a complete set of job files, shop drawings, maintenance and operating manuals, and as-built drawings.

Recent years have witnessed what can only be describe as increasing neglect of construction quality control. Modern procedures involve so many players on the construction team that responsibility for the end product has become splintered and diffused. The withdrawal of the architect-engineer has been previously mentioned. Inspection by the contractor is criticized on the basis that it is unrealistic to expect that profit-oriented contractors will rigorously inspect their own work. There seems to be increasing sentiment that construction quality control should be a budgeted part of the total construction process, and that this responsibility should be contracted out to third-party, professional firms.

17.14 Expediting

Construction is of such a nature that the timely delivery of project materials is of extreme importance. If required items are not available when needed, the contractor can experience major difficulties because of the disruption of the construction schedule. Such delays are expensive, awkward, and inconvenient, and every effort must be made to avoid them. When purchase orders are written, delivery dates are designated which, if met, will ensure that the materials will be available when needed. These dates are established on the basis of the project progress schedule and must necessarily make allowance for the approval of shop drawings.

Unfortunately, the contractor cannot assume that the designation of delivery dates in its purchase orders or the securing of delivery promises from the sellers will automatically ensure that the materials will appear on schedule. To obtain the best service possible, a series of follow-up actions, referred to as expediting, are taken after each material order is placed to keep the supplier constantly reminded of the importance of timely delivery. Expediting may be a jobsite function or the construction firm may provide all of its construction projects with a centralized expediting service. A full-time expediter is sometimes required on a large project. On work where the owner is especially concerned with job completion, or where certain material deliveries are crucial, the owner will often participate with the contractor in cooperative expediting efforts.

A necessary adjunct to the expediting function is the maintaining of a check-off system or log where the many steps in the material delivery process are recorded. Starting with the issuance of the purchase order, a record is kept of the dates of receipt of shop drawings, their submittal to the architect-engineer, receipt of approved copies, return of the approved drawings to the vendor, and delivery of the materials. Because shop drawings from subcontractors are submitted for approval through the general contractor, the check-off system will include materials being provided by the subcontractors. This is desirable because project delay can be caused by any late material delivery, regardless of who provides the material. This same documentation procedure is followed for samples, mill certificates, concrete-mix designs, and other submittal information required. General contractors sometimes find it necessary on critical material items to determine the manufacturer's production calendar, testing schedule if required, method of transportation to the site, and data concerning the carrier and shipment routing. This kind of information is especially helpful in working the production and transportation around strikes and other delays.

Each step in the approval, manufacture, and delivery process is recorded and the status of all materials is checked frequently. At intervals, a material status report is forwarded to the project manager for his information. This system enables job management to stay current on material supply information and serves as an early-warning device when slippages in delivery dates seem likely to occur.

The intensity with which the delivery status of materials is monitored depends on the nature of the materials concerned. Routine materials such as sand, gravel, brick, and lumber usually require little follow-up. Critical made-to-order items, whose late delivery would badly cripple construction operations, must be closely monitored. In such cases, the first follow-up action would be taken weeks or months in advance of the scheduled delivery date. This action, perhaps a letter showing order number, date of order, and delivery promise, would request specific information on the anticipated date of shipment. Return answers to such inquiries can be very helpful. If a delay appears forthcoming, strong and immediate action is necessary. Letters, telegrams, telephone calls, and personal visits, in that order, may be required to keep the order progressing on schedule.

17.15 Deliveries

In addition to working for the timely delivery of materials, the expediter is also usually responsible for their receipt, unloading, and storage. In general, deliveries are made directly to the projects to minimize handling, storage, insurance, and transportation costs. However, there are often instance where it is preferable or necessary to store materials temporarily at off-site locations until they are needed on the job.

When notice of a material delivery is received, suitable receiving arrangements must be made. Advance notice of shipments is provided directly by the vendors or through bills of lading or other shipping papers. If a shipment is due on a jobsite, notice is given to the project superintendent. If suitable unloading equipment is not available on the site, such equipment must be scheduled or the project superintendent be authorized to obtain whatever may be required. If a shipment is to be made to the contractor's storage yard or warehouse, the person in charge must be advised and unloading equipment scheduled if required. In the case of railcar shipments, unloading must be accomplished before expiration of the free unloading period or the car goes under demurrage. Demurrage is a daily charge made by the railroad until the car is released by the contractor.

The scheduling of material deliveries to the jobsite can be especially important on some projects. For example, consider the delivery of structural steel to a building project in a downtown city area. <u>On projects of this type, storage space is extremely limited and deliveries must be carefully scheduled to arrive in the order needed and at a rate commensurate with the advancement of the structure.</u>[9] Such projects obviously require careful scheduling of deliveries and the close cooperation of the contractor and the steel supplier. An additional factor is the routing of the trucks through the city streets, often at off-hours, and arranging for the direction of traffic around the vehicles during the delivery and unloading operations Arrangements for necessary permits, police escorts, labor, and unloading equipment must be made in advance. On such projects, many material deliveries are not made directly to the site but to temporary storage facilities owned or rented by the contractor. When such off-site storage is used, deliveries to the jobsite are made in accordance with short-term job needs.

17.16 Receiving

Delivery of job materials is usually made directly to the jobsite. There are times, however, when it is either undesirable or impossible to accept shipments at the project. Construction in congested urban areas is an instance already mentioned. Another example is early delivery of items that would be susceptible to damage, loss, or theft if stored on the job for extensive periods of time. Whenever possible, such materials are stored in contractor's yard or warehouse until they are needed.

Truck shipments may be made by common carriers or the vendor's own vehicles. In either case, the material must be checked for damage as it is being unloaded and quantities checked against the freight bill or vendor's delivery slip. Observed damage must always be noted on all copies of the freight bill and be witnessed by the truck driver's signature. The receiver told not sign the delivery slip or freight bill until he has checked the quantity delivered against that indicated.

When shipment is made by rail car, the contractor advises the carrier where it desires the car to be spotted as soon as the contractor is advised of the car number. The shipment should be checked after the car is placed for unloading and any visible damage reported to the railroad claim agent. In case of damage, unloading must be deferred until the shipment has been inspected and proper notations made on the bill of lading. A claim for damage or loss is submitted to the freight claim agent on the standard form of the carrier. This claim must be accompanied by the original bill of lading, the receipted original freight bill, the original or a certified copy of the vendor's invoice, and other information in substantiation of the claim. Should damage be such that it is not visible and cannot be detected until the goods are unpacked, the contractor must make its claim at that time on the carrier's special form which is used for concealed damage. Rail shipments of less-than-carload (LCL) may be such that the contractor must pick up the material at the freight depot, or it may be delivered to its project or yard by truck. Delivery depends on the FOB point designated by the purchase order.

The party who receives a shipment on behalf of the contractor should immediately transmit the covering delivery ticket, freight bill, or bill of lading to the contractor's office. Information pertaining to damage or shortage and location of material storage should be included.

Although most purchase orders include freight charges in their face amounts, material vendors do not always prepay freight charges. As a result, materials often arrive at the contractor's location with freight charges collect. The contractor usually has an account with the carrier that allows it to receive the goods without having to pay the transportation charges at the time of delivery. However, in the case of common carriers, Interstate Commerce Commission regulations require that freight charges be paid within a short time of delivery. Therefore, it is important that freight bills for collect shipments be transmitted immediately to the contractor's office for payment. Where the purchase order amount included freight, it is usual for the contractor to pay the freight charges and back-charge the account of the vendor.

17.17 Inspection of Materials

Inspection of delivered goods for quantity and quality should preferably be done concurrently with their unloading and storage. This is not always possible, however, and often would involve an objectionable delay in the unloading and release of transporting equipment. The checking of the package count as shown by the freight bill or delivery ticket should always be done, with any variations being indicated on the bill or ticket. However, the quantities of different items and their quality often must be verified at the first opportunity after receipt. To do this, the party making the inspection obtains copies of the covering purchase order and approved shop drawings. A thorough check of the delivered items is made to verify both item quantities and quality. This verification can become quite laborious in some cases, but can pay big dividends in minimizing later job delays because of missing, faulty, or erroneous materials. The project inspector will frequently participate and assist in this inspecting process. Inspection of material deliveries must be done with reasonable promptness so that there is time to take any corrective measures that may prove necessary.

After the delivered materials have been reconciled with the shop drawings and purchase orders, the inspector will normally file a receiving report with the company's procurement section. This report shows the purchase order number, date of inspection, material, location of storage, quantity, remarks, and signature of the inspector. The receiving report verifying receipt of the proper count and quality clears the order and authorizes payment to the vendor. With partial shipments, several receiving reports may be required to clear the entire order.

Inspection duties at times involve the sampling of various kinds of construction materials. Construction contracts may require the laboratory testing of certain materials as proof of quality. Thus, inspectors must be acquainted with the standard methods of sampling sand, gravel, bulk cement, asphalt, reinforcing steel, and other construction commodities. Another aspect of materials inspection is the obtaining of certification of quality or the results of laboratory control tests from the manufacturer or producer. Submittal of the manufacturer's certification of quality may enable the contractor to avoid duplicate acceptance testing.

17.18 Subcontractor Scheduling

As important members of the field construction team, subcontractors have an obligation to pursue their work in accordance with the project schedule established by the prime contractor. Failure of a subcontractor to commence its operations when required and to pursue its share of the work diligently can be a serious matter for the general contractor. Consequently, the scheduling of subcontractors deserves and receives appropriate action by the general contractor. A practice followed by many contractors is to notify each subcontractor by letter two weeks or more before the subcontractor is expected to move onto the project and commence operations. Subcontractors must be given adequate time to plan their work and make the necessary arrangements to start their operations. Follow-up telephone calls are made if needed. In the interest of good subcontractor relations, the project manager should not schedule a subcontractor to appear on the site until the job is ready and the subcontracted work can proceed unimpeded. After the subcontractor is on the job, its progress must be monitored to ensure its operations are keeping pace with the overall project time schedule. If its work falls behind, the project manager may reasonably instruct the subcontractor to take appropriate measures to accelerate its progress.

With regard to the general matter of subcontractor scheduling, the form and content of the subcontract can be very important. A carefully written document with specific provisions regarding conformance with time schedule, material orders, and shop drawing submittals can strengthen the project manager's hand in keeping all aspects of the project on schedule. In this context, a common problem is the failure of a subcontractor to order major materials in ample time to meet the construction schedule. General contractors occasionally find it advisable to monitor their subcontractors' material purchases. This can be accomplished by including a subcontract requirement that the subcontractor submits unpriced copies of its purchase orders to the general contractor within ten days after execution of the subcontract. In this way, the general contractor can oversee the expediting of key materials provided by subcontractors along with its own.

17.19 Record Drawings

A common general contract requirement is that the contractor must maintain and prepare one set of full-size contract drawings marked to show various kinds of "as-built" information. These drawings show the actual manner, location, and dimensions of all work as actually performed. This involves marking a set of drawings to show details of work items that were not performed exactly as they were originally shown, such as changed work, changed site conditions, and variations in alignment or location. In addition, details and exact dimensions are given for those work items that were not precisely located on the original contract drawings. Depths, locations, and routings of electrical service and underground piping and utilities are examples of this point. The set of record drawings is prepared by the contractor as the work progresses and is turned over to the architect-engineer or owner at the end of the project.

17.20 Disbursement Controls

To coordinate the actions of the company accounting office with the project it is necessary to implement a system of disbursement controls. These controls are directed toward controlling payments made to vendors and subcontractors and require that no such payments be made without proper approval from the field. The basic purpose of disbursement control is twofold; first, to ensure that payment is made only up to the value of the goods and services received to date and, second, to see that total payment does not exceed the amount established by the purchase order or subcontract.

Payments made for materials are based on the terms and conditions of covering purchase orders. Copies of all job purchase orders are provided to the project manager for his use and information. Purchase order disbursement by the accounting office is conditioned on the receipt of a signed delivery ticket or receiving report from the jobsite. Suitable internal controls are established to ensure that total payments do not exceed the purchase order amount. Any change in purchase order amount, terms, or conditions is in the form of a formal written modification, with copies sent to the jobsite.

Disbursements to subcontractors follow a similar pattern. Because there are no delivery tickets or receiving reports tot subcontractors, all subcontractor invoices are routed for approval through the project manager, who has copies of all the subcontractors. The project manager determines if the invoice reflects actual job progress and approves the invoice or makes appropriate changes. General contractors normally withhold the same percentage from their subcontractors that owners retain from them. If the subcontractor bills for materials stored on site, a common requirement is that copies of invoices be submitted to substantiate the amounts billed. Any change to a subcontract, is accomplished by a formal change order.

17.21　Job Records

To serve a variety of purposes, a documentation system is needed on each project that will produce a comprehensive record of events that transpired during the construction period. The extent to which this is done and the job records that are maintained are very much functions of the provisions of the construction contract and the size, complexity, and risks inherent in the work. It is up to the contractor to establish what records are appropriate for a given project and to see that they are properly kept and filed.

The original estimating file and the contract documents are obvious basic job records. During the construction phase, periodic progress reports, cost reports, the job log, correspondence, minutes of job meetings, time schedules, subcontracts, purchase orders, field surveys, test reports, progress photographs, shop drawings, and change orders are routinely maintained as a permanent project record. On larger projects additional records on manpower, equipment, operating tests, back charges, pile-driving and welding records, progress evaluation studies, and others are kept.

17.22　The Daily Job Log

A job log is a historical record of the daily events that take place on the jobsite. The information to be included is a matter of personal judgment, but the log should include everything relevant to the work and its performance. The date, weather conditions, job accidents, numbers of workers, and amounts of equipment should always be noted. It is advisable to indicate the numbers of workers by craft and to list the equipment items by type. A general discussion of the daily progress, including a description of the activities completed and started and an assessment of the work accomplished, is important. Where possible and appropriate, the quantities of work put into place can be included. The job log should be maintained in a hardcover, bound booklet or journal. Pages should be numbered consecutively in ink with no numbers being skipped. Every day should be reported and all entries made on the same day as they occur. No erasures should be made, and each diary entry should be signed immediately under the last line of the day's entry.

The diary should list the subcontractors who worked on the site together with the workers and equipment provided. Note should be made of the performance of subcontractors and how well they are conforming to the project time schedule. Material deliveries received must be noted together with any shortages or damage incurred. It is especially important to note when material delivery dates are not met and to record the effect of such delays on job progress and costs.

The diary should include the names of visitors to the site and facts pertinent thereto. Visits by owner representatives, the architect-engineer, safety inspectors, union representatives, and people from utilities and government agencies should be documented and described. Meetings of various groups at the jobsite should be recorded, including the names of people in attendance, problems discussed, and conclusions reached.

Complete diary information is occasionally necessary to substantiate payment for extra work and is always needed for any work that might involve a claim. The daily diary should always include a description of job problems and what steps are being taken to correct them. The job log is an especially important document where disputes result in arbitration or litigation. To be accepted by the courts as evidence, the job diary must meet several criteria. The entries in the log must be original entries made on the dates shown. The entries must have been made in the regular course of business and must constitute a regular business record. The entries must be original entries, made contemporaneously with the events being recorded and based on the personal knowledge of the person making them. Where the above criteria have been met, the courts have generally ruled that the diary itself can be entered as evidence, even if the author thereof is not available to testify.

17.23 Claims

Construction contracts typically require the contractor to advise the owner in writing, within a prescribed period of time, concerning any event that will result in delay and/or additional cost. The actual claim against the owner follows and includes a detailed description of the job condition underlying the claim, identification of the contract provisions or legal basis under which the claim is made, details of how the condition has caused the extra cost and/or delay, and a summary of the increased costs and extension of time requested. A claim usually involves both questions of entitlement which refers to the merit of the claim, and the cost and time extension involved. Claims stem from a wide variety of conditions including (1) breach of contract, (2) changes, (3) constructive changes, (4) changed conditions, (5) delay or interference, (6) acceleration, (7) errors or omissions in design, (8) suspension of the work, (9) variations in bid-item quantities, or (10) rejection of or-equal substitutions. Almost any extra cost or time caused to the contractor by the action or inaction of the owner or the owner's agent can be a valid basis for a claim against the owner. Refusal by the owner to recognize the claim does not ordinarily authorize the contractor to refuse to continue its field operations. However, the contractor should proceed with the work in dispute only after filing a written protest with the owner.

Most claims are resolved by negotiation between the owner and contractor. Otherwise, settlement is ultimately reached in the courts, by arbitration, or through contract appeal boards. Regardless of the means by which the matter is ultimately decided, the contractor often does not prevail, not because the claim has no merit but because of the contractor's inability to prove actual damages. This inadequacy is usually a direct result of the fact that the contractor did not maintain adequate records to support its position. The usual records maintained for project management are not necessarily adequate for proof of a claim.

In the final analysis, the successful settlement of a disputed claim depends largely on painstaking documentation. Preparation of a claim may be based on the routine project records compiled during the construction process. However, these records may not contain the detailed information needed to substantiate such a demand. For this reason, the contractor may be well advised to maintain a special set of records that pertain specifically to the matter in dispute. The standard dictum of "put everything in writing" applies here. It is to be noted in this regard that, to be effective documentation must be prepared during the construction process. Records created after the fact will not generally receive consideration. Project photographs, dated and identified, can be very effective when working with claims.

New Words and Expressions

unremitting   *adj.*   持续的，不间断的，不懈的
payroll   *n.*   （公司员工的）工资名单，（公司的）工资总支出
superintendent   *n.*   主管人，负责人，监督人
jurisdiction   *n.*   司法权，裁判权，权限，管辖权
agenda   *n.*   议事日程，记事册
liquidate   *v.*   清理，清算，消灭，取消
bonus   *n.*   额外津贴，奖金，红利
vendor   *n.*   卖主，小贩
minute   *n.*   会议记录，备忘录
shop drawing   施工图
wiring   *n.*   布线，装设金属线，接线图，布线图
millwork   *n.*   预制构件，工厂预制构件，细木工制品
liability   *n.*   责任，义务，负债，倾向
oversight   *n.*   失察，忽略，监督，负责
expediter   *n.*   调度员，原料供给者
demurrage   *n.*   滞留期，滞留费
freight   *n.*   货物，货运，运费
invoice   *n.*   发票，装货清单，货物的托运
depot   *n.*   仓库
dividend   *n.*   红利，股息
oversee   *v.*   检查，监督，俯瞰
record drawings   竣工图，施工记录图
disbursement   *n.*   支付，付出款，分配
transpire   *v.*   蒸发，泄漏，发生
pile-driving   打桩
thereto   *adv.*   向那里，此外
arbitration   *n.*   仲裁，公断
litigation   *n.*   诉讼，争论

entitle　*v.*　给……权利（资格），给……标题（称号）

breach　*n. & v.*　违反，不履行，突破

painstaking　*adj.*　辛勤的，艰苦的

date　*v.*　注明……的日期

 Notes

1. These are, after all, the heart and soul of any such business enterprise. If a project is to meet its established time schedule, cost budget, and quality requirements, close management control of field operations is a necessity.

本句难点解析：after all 为插入语，these 为指示代词，指代前文提到的"项目"。be to 表示"目的"。close 是一个多义词，在这里意为"密切的，紧密的"。

本句大意如下：毕竟，对于从事建筑行业的企业，项目都是它的核心和灵魂。如果一个项目要满足既定的施工进度、工程造价预算和施工质量要求，那么对现场作业进行密切的管控就是非常有必要的。

2. Project manager is a function of executive leadership and provides the cohesive force that binds together the several diverse elements into a team effort for project completion.

本句难点解析：a function of 意为"……的一种职能"，此处指的是领导职能。bind together 意为"将……结合起来"，此处为"团结"之意。

本句大意如下：项目经理具有行政领导职责，他需要提升集体的内部凝聚力，将各个有利于项目的组成部分有效地团结起来。

3. In practice, construction project authority is wielded much as a partnership effort, with the project manager and project superintendent functioning much as allied equals.

本句难点解析：functioning much as allied equals 现在分词短语做后置定语，修饰前文的项目经理与项目负责人。

本句大意如下：在实际工程中，建设项目部门很类似团队合作，而项目经理和项目负责人的关系更像是同盟。

4. Many things can demotivate a craftsman, but the CICE study reports that some of the most common complaints are: materials unavailable, unsafe working conditions, redoing work already completed, unavailability of tools or equipment, lack of communication, and disrespectful treatment by supervisors.

本句大意如下：有很多情况可能会导致工人消极怠工，但是 CICE 研究报告表明，最普遍的抱怨是：没有可用的材料、工作条件不安全、对已完成工作进行返工、工具或者设备失效、缺乏沟通或是监管人员行事无礼。

5. The efficient handling, control, and disposition of contractual and administrative matters in a timely fashion are of paramount importance for a smooth-running job.

本句难点解析：be of importance 在此处相当于 important。

本句大意如下：高效的处理、控制、及时处置合同和行政事项对于工作的顺利运行至关重要。

6. These regular meetings are very valuable in the sense that all parties concerned are kept fully informed concerning the current job status and impressed with the importance of meeting their own obligations and commitments.

本句难点解析：in the sense that 意为"从某种意义上来说"。

本句大意如下：这些日常会议是非常有价值的，可以使相关各方了解目前的工作状况，并强化履行自己的义务和责任的重要性。

7. By establishing the total costs associated with each scheduled segment of the project and by making some reasonable assumption concerning how construction cost varies with time (a linear relationship is often assumed), a reasonably accurate prediction of the value of construction in place at the end of each month can be made.

本句难点解析：by establishing 和下文的 by making 是并列的关系，通过这两种方式，来实现最后的目的。the value of construction in place 其中的 value 在此处为"进度"的意思。

本句大意如下：通过建立与工程的每个计划分段相关的总花费和通过假设施工花费随时间变化的情况（通常假设是线性关系），可以准确地预测每个月的月底现场施工的进度。

8. To illustrate, the responsibility for field inspection has led the liability of architect-engineers for construction defects caused by the contractor's failure to follow contract requirements.

本句难点解析：整个句子可以简化为 the responsibility has led the liability，其余部分均为定语，起到修饰这两个名词的作用。failure to follow，没有成功地执行，即没有满足合同要求。

本句大意如下：为对此说明，现场检查的职责为建筑工程师承担了由于承包商未满足合同要求而导致的施工缺陷的责任。

9. On projects of this type, storage space is extremely limited and deliveries must be carefully scheduled to arrive in the order needed and at a rate commensurate with the advancement of the structure.

本句难点解析：a rate commensurate with 意为"与……相称的速度"。the advancement of the structure 意为"结构的进展"，可以引申为结构的完成进度。

本句大意如下：对这种类型的项目而言，储存空间是极其有限的，应仔细制定运输计划，确保材料按所需顺序进场，并且使材料进场的速度与结构的完成进度相匹配。

Translate the following phrases into Chinese.
1. unremitting management effort
2. functional line
3. material and equipment dealers
4. project superintendent
5. disrespectful treatment by supervisors
6. architect-engineer
7. a check-off system

8. back-charge the account of the vendor

9. take appropriate measures

10. job log

Translate the following sentences into Chinese.

1. Project management and field supervision are quite different responsibilities. The day-to-day direction of project operations is handled by a site supervisor or project superintendent. His duties involve supervising and directing the trades, coordinating the subcontractors, working closely with the owner's or architect-engineer's field representative, checking daily production, and keeping the work progressing smoothly and on schedule.

2. Specifically, project administration refers to those practices and procedures, usually routine, that keep the project progressing in the desired fashion.

3. The working drawings and specifications prepared by the architect-engineer, although adequate for job pricing and general construction purposes, are not suitable for the fabrication and production of many required construction products.

4. Unfortunately, the contractor cannot assume that the designation of delivery dates in its purchase orders or the securing of delivery promises from the sellers will automatically ensure that the materials will appear on schedule.

5. Most claims are resolved by negotiation between the owner and contractor. Otherwise, settlement is ultimately reached in the courts, by arbitration, or through contract appeal boards.

Unit 18

Design and Construction of the Jin Mao Tower

The composite structural system for the Jin Mao Tower located in Shanghai, China was designed to resist typhoons, earthquake forces, and accommodate poor soil conditions while providing a very slender tower to be fully occupied for the office and hotel uses.[1] The structure for the Jin Mao Tower was completed in August 1997 with occupancy expected in August 1998. The Tower is the tallest building in China and the third tallest in the world (as of 1997). The superstructure of the Tower resists lateral loads with a central reinforced concrete shear wall core interconnected with composite mega-columns. The outrigger trusses at three (3) two-story levels. Gravity loads are resisted by composite floor members which frame to structural steel mega-columns and to the central core and composite mega-columns.[2] The foundation system for the Tower consists of open steel pipe piles capped by a reinforced concrete mat. A deep slurry wall system forms the temporary retention as well as the permanent foundation system around the site.

18.1 The structure system

The superstructure for the 421 meter-tall, 88-story Jin Mao Tower consists of a mixed use of structural steel and reinforced concrete with major structural members composed of both structural steel and reinforced concrete (composite). Thirty-six (36) stories of hotel spaces exist over 52 stories of office space. The structure is being developed by the China Shanghai Foreign Trade Co., Ltd. and constructed by the Shanghai Jin Mao Contractors, a consortium of the Shanghai Construction Group; Obayashi Corp., Tokyo; Campenon Bernard SGE, Paris; and Chevalier, Hong Kong. The structure was topped-out in August 1997 with an expected overall completion date of August 1998. The structure is the tallest in China and the third tallest in the world behind the Petronas Towers in Kuala Lumpur, Malaysia and the Sears Tower in Chicago, Illinois, USA (as of 1997).

The primary components of the lateral system for this slender Tower, with an overall aspect ratio of 7 : 1 to the top occupied floor and an overall aspect ratio of 8 : 1 to the top spire, include a central reinforced concrete core wall linked to exterior composite mega-columns by structural steel outrigger trusses.[3] The central core wall houses the primary building service functions, including elevators, mechanical fan rooms for HVAC services, and washrooms. The octagon-shaped core is nominally 27m deep with flanges varying in thickness from 850mm at the top of the foundations to 450mm at the level 87 with concrete strength varying from C60 to C40. Four (4) - 450mm thick interconnecting core web walls exist throughout the office levels with no web walls on the hotel levels, creating an atrium with a total height of 205m which leads into the spire, the composite mega-columns vary in cross-section from 1500mm×5000mm at the top of foundations to 1000mm × 3500mm at Level 87. Concrete strengths vary from C60 at the lowest floors to C40 at the highest floors.

Structural steel outrigger trusses interconnect the central core and the composite mega-column at three 2 - story tall levels. The interconnection occurs between Level 24 & Level 26, Level 51&

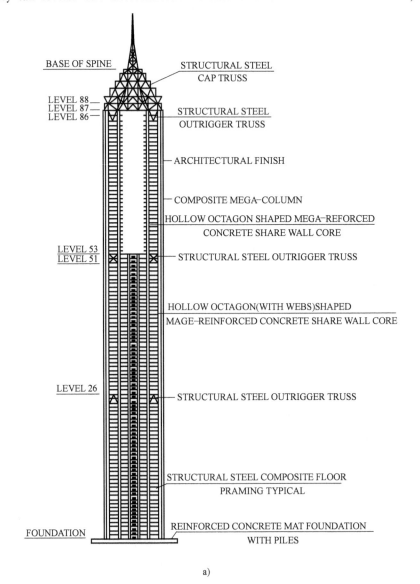

a)

Fig. 18-1 Structure system elevation and framing plans.

a) Tower structural system elevation

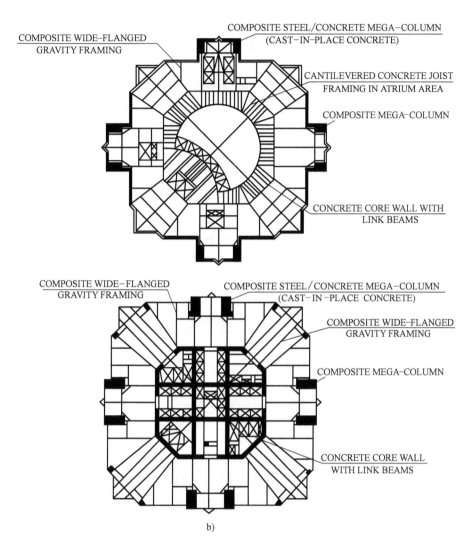

Fig. 18-1 Structure system elevation and framing plans. (Continued)
b) Typical office framing plan

Level 53, and Level 85 & Level 87. The outrigger trusses between Level 85 & Level 87 engage the 3-dimensional structural steel cap truss system. The cap truss system which frames the top of the building between Level 87 and spire is used to span over the open core, support the gravity load of heavy mechanical spaces, engage the structural steel spire, and resist lateral loads above the top of the central core wall/composite mega-column system.[4]

In addition to resisting lateral loads, the central reinforced concrete core wall and the composite mega-columns carry gravity loads. Eight (8) built-up structural steel mega-columns also carry gravity loads and composite structural steel wide-flanged beams and built-up trusses are used to frame typical floors. The floor framing elements are typically spaced at 4.5m on-center with a composite metal deck slab (75mm metal deck topped with 80mm of normal weight concrete) framing between the steel members. Fig. 18-1 illustrates the components of the superstructure.

18.2 Foundation engineering

Because of the extremely poor upper-strata soil conditions, deep, high-capacity structural steel pipe piles are required to transfer the superstructure loads to the soil by friction. Open structure steel pipe piles are 65m long with a tip elevation 80m from existing grade. The tips of the pipes rest in very stiff sand and are the deepest ever attempted in China. Pipe piles were installed in three (3) approximately equal segments, having a wall thickness of 20mm, and having an individual design pile capacity of 750 tones. Piles were driven from grade with 15m long followers before any site retention system construction or excavation had commenced. The pipe piles are typically spaced at 2.7m on-center under the core and composite mega-columns with a 3.0m spacing under the other areas. The piles are capped with a 4m thick reinforced concrete mat comprised of 13500m^3 of C50 concrete. The mat was poured continuously, without any cold joints, over a 48 hour period. Concrete temperature was controlled by an internal cooling pipe system with insulating straw blankets used on the top surface to control the temperature variations through the depth of the mat and to control cracking.[5]

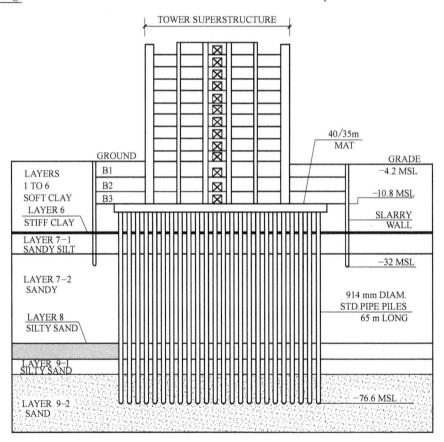

Fig. 18-2 Tower foundation systems.

A reinforced concrete slurry system was designed and constructed around the entire perimeter of the site (0.75kilometer). The thickness of the slurry wall is 1 m with a concrete design strength of C40 and depth of 33m. The slurry wall bears on moderately stiff, impervious clay, also acting as a temporary retention system wall, a permanent water cut-off system. A tieback ground anchor system was designed and successfully tested to provide lateral support of the slurry wall during construction, however, the contractor chose to construct a locally accepted reinforced concrete cross-lot bracing system for the three (3) full basement levels which extended approximately 15m below grade. The permanent ground water table is within 1 m of existing grade. Based on the site conditions and the slurry wall depth, a sub-soil draining system was designed to carry 18.5liter/sec of water. An overall description of the foundation system is shown in Fig. 18-2.

18.3 Wind engineering

抗风设计

Typhoons as well as strong extratropical winds exist in the local Shanghai environment. Multiple analytical and physical testing techniques were used to evaluate the behavior of the Tower. Since ultra-tall structures had not been previously constructed in China, the Chinese wind design code did not address structures taller than 160m. Therefore, code requirements were extrapolated for the Tower and wind tunnel studies were performed to confirm Code extrapolations and to study the actual, "rational" local wind climate. Wind tunnel studies, performed under the direction of Dr. Nichol Isyumov at the University of Western Ontario in conjunction with the Shanghai Climate Center, were conducted for the building located in the existing site condition and considering the future master plan development termed the "developed Pudong" condition. The existing site context essentially consisted of low-rise building (3~5 stories in height) with the fully "developed Pudong" environment consisting of 30~50 stories buildings surrounding the Jin Mao Tower with two (2) ultra-tall towers located within 300m of Jin Mao. <u>Wind tunnel investigation included a local climate study, construction of proximity models, a force balance test, an aeroelastic test, an exterior pressure test, and a pedestrian level wind study.</u>[6] All tests considered both typhoon and extratropical winds as well as the existing and "developed Pudong" site conditions.

The final design of the Tower considered both the People's Republic of China Building Code as well as the "rational" wind tunnel studies. Strength design for all lateral load-resisting components is based on the Code-defined 100-year return wind with a basic wind speed of 33m/s for a 10 minute average time at 10m above grade. The basic wind speed corresponds to a design wind pressure for the Tower of approximately 0.7kPa at the bottom of the building and 3.5kPa at the top of the building. Results from the wind tunnel studies, considering the existing site condition and the "developed Pudong" condition as well as extratropical and typhoons confirmed that the Chinese Code requirements for design were conservative.

Serviceability design, including the evaluation of building drift and acceleration, was based on the "rational" wind tunnel study results. Wind tunnel studies were performed for 1-year, 10-year, 30-year, 50-year and 100-year return periods. The studies considered the actual characteristics of the structure. The fundamental translational periods of the structure are 5.7 seconds in each principal direction and the fundamental torsional period is 2.5 seconds. The overall building drift, with comparable inter-story drifts, for the 50-year return wind with 2.5% structural damping is $H/1142$ for the existing site condition and $H/857$ for the "developed Pudong" condition. It was determined that the two (2) ultra-tall structures proposed to be located near the Jin Mao Tower would have a significant effect on the dynamic behavior resulting in significantly higher effective structural design pressures. Building drifts are well within the internationally accepted building drift of $H/500$. Considering 1.5% structural damping and a 10-year return period, the expected building acceleration ranged from $9 \sim 13$ milli-g's for the top floor of the occupied hotel zone. In addition, expected building acceleration ranged from $3 \sim 5$ milli-g's for a 1-year return period considering 1.5% structural damping. The internationally acceptable accelerations for a hotel structure are $15 \sim 20$ milli-g's for a 10-year return period and $7 \sim 10$ milli-g's for a 1-year return period. Because of the favorable serviceability behavior of the building, the passive characteristics alone could be used to control dynamic behavior with no additional mechanical damping required.

Wind tunnel study results determined that the Code requirements for lateral load design was equivalent to a 3000-year return wind. The overall building drift based on this conservative wind loading is $H/575$ which also meets internationally accepted standards for drift.

18.4　Earthquake engineering

The approach for evaluating seismic loadings for the Jin Mao Tower considers both Chinese Code-defined seismic criteria and actual site-specific geological, tectonic, seismological and soil characteristics. Actual on-site field sampling of the soil strata and engineering evaluations were performed by Woodward-Clyde Consultants, the Shanghai Institute of Geotechnical Investigation and Surveying, and the Shanghai Seismological Bureau.

All lateral load resisting systems, including all individual members, were designed to accommodate forces generated from the Chinese Code-defined response spectrum as well as site specific response spectrums.[7] Extreme event site-specific time history acceleration records (10% probability of occurrence in a 100-year return period) were used in time history analyses to study the dynamic behavior of key structural elements including the composite mega-columns, the central core, and the outrigger trusses.

The site specific response spectrums used to describe the Tower's dynamic behavior included analyses for a most probable earthquake with a 63% probability of occurrence in a 50-year return period and a most credible earthquake with a 10% probability of occurrence in a 100-year return period. In addition, the Tower was evaluated using a 3-dimensional dynamic time history analysis for a most credible earthquake with a 10% probability of occurrence in a 100-year return period.

In all cases, the Chinese-defined code wind requirements governed the overall building behavior and strength design; however, special considerations were given to the outrigger trusses and their connections. In all design cases, these structural

抗震设计

steel trusses were designed to remain elastic.

18.5 Unique structural engineering solutions/On-site structure monitoring

The structural design for the Jin Mao Tower created an opportunity to develop unique structural engineering solutions. These solutions included the practical development of the theoretical concepts, unusual detailing of large structural building components, and comprehensive monitoring of the in-place structure.[8]

The overall structural system utilizes fundamental physics to resist lateral loads. The slender cantilevering reinforced concrete central core is braced by the outrigger trusses which act as levers to engage perimeter composite mega-columns, maximizing the overall structural depth. The overall structural redundancy is limited by engaging only for four (4) composite mega-columns in each primary direction. Structural materials are strategically placed to balance the applied lateral loads with forces due to gravity. Very little structural material premiums were realized because of the structural system used. Lateral system premiums essentially related to material required for the outrigger trusses only without measurable structural material premiums required for central core wall and composite mega-columns elements.[9] The combination of structural elements provides a structural system with 75% cantilever efficiency.

Even after equalizing the stress level within the central core and composite mega-columns, the expected relative shortening between the interconnected central core and composite mega-columns was large. By calculation, considering long-term creep, shrinkage, and elastic shortening, the expected relative movement between these elements at Levels 24~26 was as much as 50mm. The magnitude of relative movement would have induced extremely high stresses into the stiff outrigger truss members weighing as much as 3280 kg/m. Therefore, structural steel pins with diameters up to 250mm were detailed into the outrigger truss system (see Fig. 18-3). These pins were installed into circular holes in horizontal members and slots in diagonal members to allow the outrigger trusses to act as free moving mechanisms for a long period during construction. This allowed a majority of the relative movement to occur free of restraint, therefore, free of stress. After a long period of time, high strength bolts were installed into the outrigger truss connections for the final service condition of the lateral load resisting system. The expected relative movement after the final bolting was performed as a maximum of 15mm at Levels 24~26. Considering the flexibility of the long composite mega-columns, the final forces attracted to the trusses did not appreciably increase the member and connection sizes.

一些特殊方案

A comprehensive structural survey and monitoring program was designed and implemented into the Jin Mao Tower. The mat foundation system under the Tower was initially surveyed just after pour completion in October 1995 and is currently still being surveyed. Based on a substructure / soil analysis, the expected maximum long-term Tower mat settlement is 75mm. The latest Tower mat settlement is shown in Fig. 18-4. In addition to surveying the mat for long-term settlement, the super-

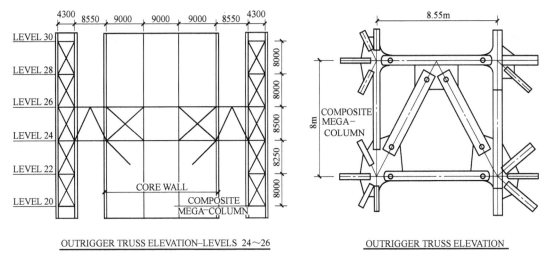

Fig. 18-3 Elevation and detail of outrigger truss system.

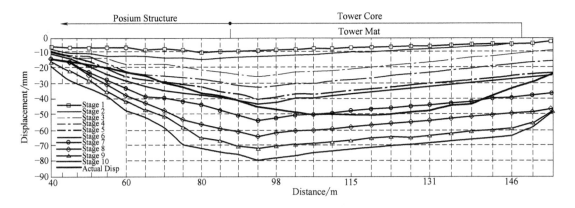

Fig. 18-4 Comparison of estimated versus actual tower mat settlement based on construction sequence.

structure was gauged to monitor behavior. Extensometers were placed on the reinforced concrete central core and on the reinforced concrete of the composite mega-columns. In addition, strain gauges were placed on the built-up structural steel mega-columns as well as on the wide flanged structural steel columns location within the concrete encasement for the composite mega-columns. Sample results of measured strain versus calculated strain are shown in Fig. 18-5. Based on the measured data, correlations were made between theoretical and actual long-term shortening of vertical elements due to creep, shrinkage, and elastic shortening. See Fig. 18-6 for results of the shear wall. Laser surveying techniques were used for both lateral and vertical building alignment. Floor levels of the structure were typically built to drawing design elevation, compensating for creep, shrinkage, and elastic shortening which occurred during construction.[10] Lateral position of the Tower was constantly monitored from off-site benchmarks and was found to be well within acceptable tolerances.

Unit 18　Design and Construction of the Jin Mao Tower

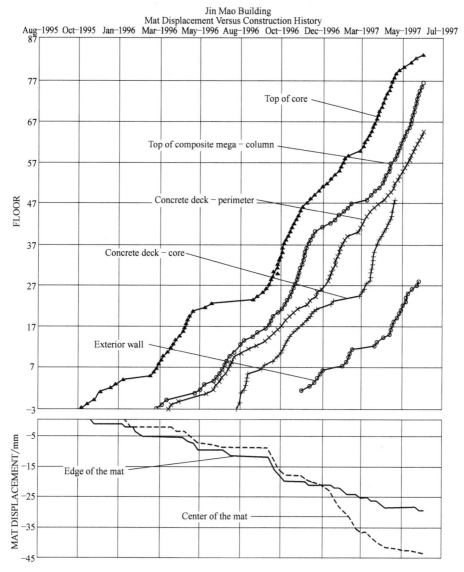

Fig. 18-5　Comparison of measured strain versus predicted strain in share wall (Level 8).

18.6　Conclusions

Incorporating fundamental structural engineering concepts into the final design of the Jin Mao Tower lead to a solution which not only addressed the adverse site conditions but also provided an efficient final design. The final structural quantities included the following for the Tower superstructure from the top of the foundation to the top of the spire (gross framed area = 205000m^2):

　　Structural Concrete　　　0.37m^3/m^2
　　Reinforcing Steel　　　　30.4kg/m^2
　　Structural Steel　　　　　73.2kg/m^2

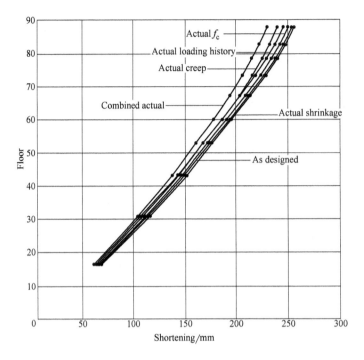

Fig. 18-6　Long-term shortening of shear wall.

A final evaluation of monitoring and survey data will be performed. Data from the as-built structure subjected to actual imposed loads will be correlated with theoretical results. This comparison will prove to be invaluable for the future design and construction of ultra-tall occupied structures.

New Words and Expressions

consortium　　*n.*　　合作，合伙，联合
topped-out　　*n.*　　封顶，上部结构
spire　　*n.*　　塔尖
octagon　　*n.*　　八边形
atrium　　*n.*　　天井，中庭
outrigger　　*n.*　　悬臂，外伸，斜撑
strata　　*n.*　　地层，岩层
grade　　*n.*　　地面
follower　　*n.*　　送桩，导架
insulating　　*adj.*　　绝缘的，绝热的
slurry　　*n.*　　稀浆，稀水泥浆
impervious　　*adj.*　　不可渗透的，透不过的
tieback　　*n.*　　拉条，拉杆，牵索
typhoon　　*n.*　　台风
wind tunnel　　风洞

Unit 18　Design and Construction of the Jin Mao Tower

redundancy　　　*n*.　　超静定，冗余
extensometer　　*n*.　　应变计
benchmark　　　*n*.　　水准基点，基础点

1. The composite structural system for the Jin Mao Tower located in Shanghai, China was designed to resist typhoons, earthquake forces, and accommodate poor soil conditions while providing a very slender tower to be fully occupied for the office and hotel uses.

本句难点解析：while 为连词，意为"同时"。句子虽然很长，但是中间基本为并列的名词和动词。

本句大意如下：中国上海金茂大厦的组合结构体系，是依据抗台风、抗震、软弱地基条件来设计的，形成一个非常细长的塔，供办公和旅馆使用。

2. Gravity loads are resisted by composite floor members which frame to structural steel mega-columns and to the central core and composite mega-columns.

本句难点解析：这里专业名词较多，structural steel mega-columns 指"型钢巨柱"；the central core 指"核心筒"，在建筑的中央部分，由电梯井道、楼梯、通风井、电缆井、公共卫生间、部分设备间围护形成中央核心筒；composite mega-column 指"组合巨柱"。

本句大意如下：重力荷载由组合楼板抵抗，其与型钢巨柱、核心筒以及组合巨柱构成整体框架。

3. The primary components of the lateral system for this slender Tower, with an overall aspect ratio of 7∶1 to the top occupied floor and an overall aspect ratio of 8∶1 to the top spire, include a central reinforced concrete core wall linked to exterior composite mega-columns by structural steel outrigger trusses.

本句难点解析：中间的 with... 为插入语，句子主干为 The components include a core wall。另外，该句的专业名词也较多，aspect ratio 为"高宽比"，outrigger trusses 为"伸臂桁架"。

本句大意如下：该塔高宽比为 7∶1（计算到最高使用楼层）和 8∶1（计算到塔尖），它的横向体系主要由钢筋混凝土剪力墙核心筒和巨型柱通过伸臂桁架连接而组成。

4. The cap truss system which frames the top of the building between Level 87 and spire is used to span over the open core, support the gravity load of heavy mechanical spaces, engage the structural steel spire, and resist lateral loads above the top of the central core wall/composite mega-column system.

本句难点解析：The cap truss system 指"帽带桁架系统"，设置在内筒和外柱间的桁架或大梁，自身具备很大的刚度，可减少内筒和外柱间的竖向变形差。which 引导定语从句，句子主干为 The cap truss system is used to span over the open core...。后面的 support、engage、resist 和 span 均为并列关系。

本句大意如下：在结构顶端（在87层和塔尖之间）形成框架的帽带桁架横跨核心筒，可承载重型机械设备间荷载，参与构成钢结构塔尖，并且抵抗核心筒和巨型复合柱顶部的水

平向荷载。

5. Concrete temperature was controlled by an internal cooling pipe system with insulating straw blankets used on the top surface to control the temperature variations through the depth of the mat and to control cracking.

本句难点解析：internal cooling pipe system 指"内部冷却管系统"。with 及其后面的内容均为修饰 pipe system 的定语。

本句大意如下：混凝土温度由内部冷却管道系统控制，顶部表面使用绝缘垫，通过垫子厚度控制温度变化并控制开裂。

6. Wind tunnel investigation included a local climate study, construction of proximity models, a force balance test, an aeroelastic test, an exterior pressure test, and a pedestrian level wind study.

本句难点解析：wind tunnel investigation 为专业名词，意为"风洞试验"。pedestrian level wind 为人行高度风。

本句大意如下：风洞试验调查研究包括当地气候研究，近似模型试验，力平衡测试，气弹模型试验，外部压力试验和人行高度风研究。

7. All lateral load resisting systems, including all individual members, were designed to accommodate forces generated from the Chinese Code-defined response spectrum as well as site specific response spectrums.

本句难点解析：response spectrum 指"反应谱"，在给定的地震加速度作用期间内，单质点体系的最大位移反应、速度反应和加速度反应随质点自振周期变化的曲线。generated from 及其后面的内容为修饰宾语 forces 的定语。

本句大意如下：所有水平向受力体系，包括所有的单个构件，依据中国规范定义的反应谱以及场地特征反应谱设计。

8. These solutions included the practical development of the theoretical concepts, unusual detailing of large structural building components, and comprehensive monitoring of the in-place structure.

本句难点解析：in-place 意为"就地，现场"。本句中为三个短语的并列，每一个短语都带有一个'of'的修饰。

本句大意如下：这些解决方案包括理论概念的发展、大型结构建筑单元的特殊构造和现场结构的综合检测。

9. Lateral system premiums essentially related to material required for the outrigger trusses only without measurable structural material premiums required for central core wall and composite mega-columns elements.

本句难点解析：without 为介词，后面跟随的为宾语 structural material premiums 和其后 required 过去分词引导的定语。

本句大意如下：水平向体系的额外费用基本上只与伸臂桁架的材料需求有关，与测量到的核心筒和组合巨柱单元所需要的结构材料的费用无关。

10. Floor levels of the structure were typically built to drawing design elevation, compensating for creep, shrinkage, and elastic shortening which occurred during construction.

本句难点解析：draw 意为"拖，拉，提取移动"。compensating for 及其后面的内容为现在分词引导的目的状语。creep 为专业名词，是指物体在荷载作用下，随时间增长而增加的变形，即徐变。

本句大意如下：为补偿施工中因徐变、收缩和弹性压缩造成的结构高度变化，结构楼层高度施工中通常要达到设计高程。

Translate the following phrases into Chinese.
1. Kuala Lumpur
2. structural steel outrigger trusses
3. resist lateral loads
4. permanent water cut-off system
5. extratropical winds
6. Chinese-defined code
7. cantilever efficiency
8. creep and shrinkage shortening
9. incorporating fundamental structural engineering concept
10. ultra-tall occupied structures

Translate the following sentences into Chinese.

1. The superstructure for the 421 meter-tall, 88-story Jin Mao Tower consists of a mixed use of structural steel and reinforced concrete with major structural members composed of both structural steel and reinforced concrete (composite).

2. Typhoons as well as strong extratropical winds exist in the local Shanghai environment. Multiple analytical and physical testing techniques were used to evaluate the behavior of the Tower.

3. In all cases, the Chinese-defined code wind requirements governed the overall building behavior and strength design; however, special considerations were given to the outrigger trusses and their connections. In all design cases, these structural steel trusses were designed to remain elastic.

4. The structural design for the Jin Mao Tower created an opportunity to develop unique structural engineering solutions. These solutions included the practical development of the theoretical concepts, unusual detailing of large structural building components, and comprehensive monitoring of the in-place structure.

5. In addition, strain gauges were placed on the built-up structural steel mega-columns as well as on the wide flanged structural steel columns location within the concrete encasement for the composite mega-columns.

Part 5

Scientific English Writing Skills

Unit 19

Scientific English Writing Skills

19.1 科技英语的基本特点

科技英语总体特点

19.1.1 总体特点

科技英语是英语语言诸变体中的一种。它与通常的文字语言表述有明显的不同。它的特点在于客观、直观、简练、准确。科技英语写作内容分类明确、特征突出，须有较好的英语基础，并在掌握了基本写作技巧后，才可以写出内容正确、文字流畅、表达准确的科技英语文章。

19.1.2 语法特点

1. 多用被动结构

【Example】 On the other hand, if the construction of buildings cannot be efficiently organized, it might result in personal or facilities injury or in destruction of equipment. Therefore, great care should be taken during the construction of buildings.

2. 多用动词非谓语结构

非谓语动词形式是指分词、动词不定式和动名词，土木工程专业文献中常用非谓语动词形式有如下两个原因：

（1）非谓语动词形式能使语言结构紧凑，行文简练

【Example】 This was the first in a series of significant developments in the structural forms of concrete high-rise buildings, freeing them from the previous 20 to 25 story height limitations of the rigid-frame and flat plate systems.

例中的 freeing them from 分词短语，如果不用非谓语形式，只能用 which make them be free from 的从句形式，那样会使语句冗长，不符合土木工程专业文献的行文要求。该句是现在分词做定语的例子。除了现在分词，英语中的非谓语动词还有过去分词、动名词和动词不定式以及由它们组成的短语。这些非谓语动词可以在句子中担任主语、定语、状语、宾语、宾语补足语等，以相对简短的篇幅代替相应从句所表达的内容。

（2）非谓语动词形式能体现或区分出句中信息的重要程度

【Example】 High buildings using bracing systems are used to resist lateral loads.

例中谓语动词 are used 表达了主要信息，现在分词短语 using bracing systems 提供细节，即非重要信息。

3. 多用长句

土木工程科学研究的目的是揭示土木工程设计及建造的规律并解释其特点及应用。这样的工作是一个复杂的程序，而且程序间的各个环节联系紧密，为了能准确、清晰地表达这种复杂现象及其之间紧密的关系，其专业文献需要用各种不同的主从复合句，而且会出现复句中从句套从句的现象。

4. 静态倾向显著

英语倾向于多用名词，因而叙述呈静态（static）；汉语倾向于多用动词，因而叙述呈动态（dynamic）。静态倾向在科技英文中表现得十分明显，主要表现为以下几个方面：

（1）形容词或动词的名词化

主要指用名词来表达原来由动词或形容词所表达的概念，如用抽象名词来表达动作、行为、变换、状态、品质、情感等概念。名词化往往可以使表达比较简洁，结构紧凑。

（2）名词连用

名词连用是指在短语中，前面的名词作为形容词修饰最后一个名词。这种名词连用的短语结构简单，表达方便，词数少而信息量大。

使用荷载：service load。

设计规格：design specification。

材料性能：material properties。

名词连用尽量略去虚词和其他次要的词，加强了英语的名词优势，也反映了英语追求简洁的趋势。这正是以叙事和推论为特点的科技英语这种文体所需要的。

另外，在名词和其他词的选择中，有些原则希望引起读者的注意。

（1）能用名词做定语的不用动名词做定语。如：

measuring accuracy→measurement accuracy

（2）可直接用名词或名词短语做定语的情况下，要少用 of 句型。如：

accuracy of measurement→measurement accuracy

（3）若有相应的形容词形式，要尽量用形容词做定语。如：

experiment results→experimental results

5. 科技英语更注重语法的地道、准确

众所周知，汉语的表达习惯更注重词序，以及词与短语之间的内在逻辑关系，而英语常常使用各种连词、关系代词和关系副词来表示分句，以及主句与从句之间的各种关系。所以我们在撰写英语文章时一定要根据具体情况先分析句子成分，把"形式上无逻辑关系"的汉语句子成分"划归"成英语相应的句子成分，利用从句、非谓语动词，并适当添加相应的代词、连词和关系副词等，再适当巧妙运用如前所述的名词化等原则，写出地道、准确的英语。

19.1.3 词汇特点

1. 多用语体正式的词汇

英语中有许多意义灵活的动词短语，而土木工程专业英语文献中则多用与之对应的意义明确的单个词构成的动词，如用 absorb 代替 take in，用 discover 代替 find out。这类动词除了具有意义明确、精炼的特点外，还具有语体庄重、正式的特点。下面是一组对照，横线右边的词汇更适合于土木工程专业英语文献。

to use up—to exhaust   to take up, to take in—to absorb
to push into—to insert   to speed up—to accelerate
to increase in amount—to accumulate   to throw bake—to reflect
to put in—to add   to keep up—to maintain
to pull towards—to attract   to carry out—to perform
to use up—to consume   to breathe in—to inhale
to think about—to consider   to throw out, to get rid of—to eliminate
to find out—to discover   to push away—to repel
to take away—to remove   to get together—to concentrate
to pour out over the top—to overflow   to fill up—to occupy
to drive forward—to propel   to pass on—to transmit, to transfer

同时，单个的英语词汇也有正式和非正式之分，土木工程专业英语文献属于正式文体，因此多用正式词汇。下面是一组对照，横线左边为非正式用语，横线右边为正式用语。

carry—bear   finish—complete   oversee—supervise
underwater—submarine   hide—conceal   buy—purchase
enough—sufficient   similar—identical   inner—interior
handbook—manual   careful—cautious   help—assist
try—attempt   feed—nourish   stop—cease
get—obtain   deep—profound   leave—depart
about—approximately   use—employ/utilize

2. 科技英语词汇的构成

随着科学技术的发展、研究的深入，科技词汇的队伍也不断壮大，各个学科都建立了自己的专业词汇库。科技文章中除了普通词汇，还有半专业词汇、专业词汇等。

（1）半专业词汇

半专业词汇的技术意义常常与其非技术意义不同，并且在不同的技术领域，半专业词汇的含义也不尽相同。所以，半专业词汇常常出现常用词汇专业化及同一词语词义多专业化的现象。例如：

concrete 具体→混凝土   angle 角度→角钢
reinforcement 加强→配筋，钢筋   channel 槽道→槽钢
detail 细节→构造，细部   web 网→腹板
stirrup 马蹬→箍筋   foot 脚→支座
cap 帽子→承台，盖梁   coat 外套→涂装（动词）
frame 骨架→框架   aggregate 聚集→集料，骨料
beam 光束→梁   bay 海湾→开间
joint 连接，关节→节点，伸缩沉降缝，后浇带   pile 堆，堆起→桩
support 支撑→支座   anchorage 停泊处→预应力的锚具
bearing 轴承→桥梁中的支座   butt 顶撞，碰撞→对接（焊缝）

（2）专业词汇

专业词汇的语义精确，所指范围小。专业词汇采用了英语构词法，大量新词随着科技的发展不断产生。了解构词法在科技英语中的应用规律，对我们快速而准确理解词义具有重大作用。英语的基本词汇并不多，但有很大一部分构成型词汇，即通过构词法而形成的词汇。科技英语词汇常采用以下几种构词方法。

1）复合词。复合词是两个或两个以上的旧词合成一个新词，例如复合名词、复合形容词、复合动词等。例如：

simply-supported 简支的   dust-tight 防尘的
earth-moving machine 推土机   earth-fill 填土
load-bearing capacity 承载力   pipeline 管线
high-rise building 高层建筑   water-proof 防水

2）派生词。派生词是指利用词的前缀与后缀作为词素构成新词。

前缀的特点：加前缀构成的新词只改变词义，不改变词类。

例：multistory 多层，superstructure 上部结构，substructure 下部结构，subsoil 地基土，antiseismic 抗震的，semidiameter 半径。

后缀的特点：加后缀构成的新词可能改变也可能不改变词义，但一定改变词类。我们可以从一个词的后缀判别它的词类。如：

compression $n.$ = compress + -ion（compress 是动词，-ion 是名词后缀）
liquidize $v.$ = liquid + -ize（liquid 是名词，-ize 是动词后缀）

3）缩略词。缩略词是将较长的英语单词取其首部或者主干构成与原词同义的短单词，或者将组成词汇短语的各个单词首字母拼成一个大写字母的字符串。

例：Fig. = Figure
lab. = laboratory
RC = reinforced concrete
PC = prestressed concrete
CAD = computer aided design
hyd. = hydrostatics
KWIC = key-word-in-context
TS = tensile strength
FEM = Finite element method

19.2 科技论文的组成

科技论文的写作有一定的模式，常见的在正式期刊上发表的文章一般包括如下内容：①论文标题；②作者姓名及单位；③摘要；④关键词；⑤引言；⑥正文；⑦结论；⑧致谢；⑨参考文献等。

19.2.1 标题

标题应写得简明扼要，同时又要写得能够吸引人，不要写得太笼统，另外字数不应写得太多。

在写论文标题时需要特别注意的是，应该使用名词短语，不该使用一个句子或动词不定式短语；一般也不用介词短语等形式，同时在标题中也不出现从句。这样就使标题简短明了，更能突出主题（这与科普文章、新闻报道、广告等是有所不同的）。

在少数情况下，也有用介词 on 引出的介词短语的，表示"论（关于）……"之意。如：

On the Possibility of Rejecting Certain Modes in VLF Propagation

另外，标题中开头的冠词可以省去（在写作者的通讯地址时，其组织机构名称前的定冠词也常省去；英美国家大多科技杂志名称前一般都不加定冠词）。关于标题中的大小写字母表示法通常有以下三种形式：

1）开头第一个字母大写、专有名词大写，其余均采用小写字母，如：

A fast bit-synchronization system in data transmission for HF channels

2）开头的字母及每个实词以及多于五个字母的介词、连词的第一个字母均大写，如：

Study of a Fast Frequency Searching Technique

3）全部字母均大写（这种形式一般用于计算机检索系统，但使读者认读很费劲），如：

A DISCUSSION ON SELF-ADAPTIVE SYSTEMS

目前前两种使用得最广泛，不过有些国外专家主张采用第一种形式，认为这种形式可读性比较好。

19.2.2 作者的姓名及单位

作者的姓名一定要用全称（full name），这主要是针对外国人来说的，因为他们的姓名一般是由三部分（即：教名——Christian name, first name, given name 或 forename；中间名——middle name；姓——surname, family name 或 last name）构成的，有的只有两部分，有的可由四部分构成。中国人的姓名应根据我国规定按汉语顺序用汉语拼音文字写出，如："李伟东"应写成 Li Weidong。若为了防止外国人产生误解，可以在姓的下面画一横线，或姓全用大写字母，或在姓后用逗号与名分开，使国外杂志的编辑人员知道这是你的姓。也有不少人仍按外国人的姓名顺序写法表示成 Weidong Li, Wei-dong Li, 或者写成 Wei D. Li 或 W. D. Li 的。

有些杂志把工作单位直接标在作者姓名下面，如：

Department of Information Engineering, Xidian University, Xi'an 710071, China

也有不少杂志把工作单位写在论文第 1 页的脚注中，如：

The authors are with （或 at）the Department of Electrical Engineering and Computer Sciences and the Electronics Research Laboratory, University of California, Berkeley, CA 94720.

The author is with the Department of Electrical Engineering, University of Toronto, Toronto, Ont., Canada M5S 1A4.

19.2.3 摘要

摘要有短有长，短的只有一句话，长的有好几段，但一般来说在杂志上发表的论文的摘要不宜过长。摘要通常有 100~150 个词，更确切地说，约为原文长度的 1%~5%（有的杂志规定摘要平均为全文的 3%~5%）。据说美国的一些高等学校规定，硕士论文提要以 250

个词左右为宜，博士论文提要以 350 个词左右为宜，会议论文的提要一般规定 300~500 个词或 1000 个印刷符号。

论文摘要同样要写得简明扼要，多数情况下只有一段内容，这一段应具有统一性、逻辑性、连贯性，同时还应突出重点，使它能够吸引读者，要达到使读者看了文摘后很想阅读论文的程度。一篇写得比较好的文摘应包括以下几个方面的内容：进行本项目的原因、自己做了哪些工作、得到了什么样的明显成果及其意义，不过也有人把上述某些内容放到正文的引言之中。

有人对文摘的语言结构进行分析研究后得出了文摘写作的三个突出的特点：

1）谓语动词很简单而句子的其余成分却十分复杂。

2）大量使用 be（主要用来陈述定义或表示"什么是什么"的陈述句；据说在所有科技英语的陈述句中，约有三分之一是 be 的变化形式做谓语动词）和 have（主要用来叙述事物具有某种或某些特征）的变化形式做谓语动词。

3）使用不涉及人的陈述句。

另外，文摘的主题句一般为"本文研究了（提出了、给出了）……"，英文句型可用主动句也可用被动句（被动句多见一些，这时句尾一般不必加"in this paper"），均用一般现在时。常见的句型为：

This paper deals with/discusses/studies/presents/gives/proposes/develops…

又如："本文首先讨论了……然后……"可采用这样的句型：

This paper begins with a discussion of…, followed by…

19.2.4 关键词

在文摘下面，不少国内外杂志要求写出 key words（关键词）或 index terms（索引词）或 subject terms（主题词），一般应写出 3~8 个关键词，主要用于计算机检索，一般学报也要求写出关键词。

19.2.5 引言

引言（Introduction），又称为前言。在长篇论文中也称为导论、导言、绪言、绪论等。如果说摘要是全篇论文的缩影，那么引言则是科技论文的帽子。它向读者初步介绍文章内容，解释文章的主题、目的和总纲。短篇论文通常在正文之前用简洁的语言叙述总纲，而对长篇论文和涉及某些不常见的内容的论文来说，引言的初步介绍能使读者便于阅读文章，引导读者明确地领会科学成果的意义、实验采用的方法和论文展开论点的计划等。

引言一般至少应包括下面三个内容：

1）说明论文主题和目的。

2）说明论文写作的情况及背景。

3）概述达到理想答案的方法。

需要指出的是引言不应重述摘要和解释摘要，不对试验的理论、方法和结果作详尽叙述，也不提前给出结论和建议。在引言中引用的参考文献要仔细加以选择，以便提供最重要的背景材料。同时还应根据读者对象的层次和特点以及他们的需要在写作引言时决定写作基调，如哪些术语需要对其明确定义，哪些试验程序应对其加以说明，总之，写作应做到有的放矢。在引言中，作者不要对自己的研究工作和自己的能力表示谦虚，可让读者对论文做出自己的评价，如要有这方面的内容可在文章结论的结尾时表示。

引言可以由一段文字组成也可由几段文字组成，这主要取决于论文的长短。在有的引言之前，作者常用 Introduction 来标明，而在有的引言（如只是一段文字的引言）之前往往没有使用 Introduction 来标明。

下面介绍几种引言的写作方法。

1. 直述主题开头法

直述主题开头法也就是在引言的第一句话中，直接点透论文的主题，然后再开始叙述其他因素。

例1：

Writing is one of the most difficult tasks for language learners. It is very much the case in learning English. Many students who have been leaning English for many years are very poor in writing at the sentence level. Some of the compositions are often even not revisable, because the sentences written do not make any sense. On the other hand, few students can write well in English. Why? A survey on students' study habits shows that habits of learning really make a difference in students' writing proficiency. Those who can write good English usually have good study habits, and good reading habits in particular. They pay much more attention to ideas while reading.

Language learning is a thinking process. Thinking is involved in every task of learning from reading to speaking, and especially writing. Good writing needs creative thinking. So to develop students' thinking is very important in helping students to write. And this should be done through the process of learning. This paper discusses four areas of suggestions on developing students' thinking.

在例1中，引言的第一句话就介绍了论文的主题。然后围绕写作是语言学习者最困难的任务之一介绍了本文的主要内容，对写作困难的原因进行了分析，并就此提出了如何发展学生的思维能力，为论文的全面展开进行了铺垫。这是一个典型的使用直叙主题开头的引言，这种方法最为常用。

2. 利用一个问题或一系列问题开头的方法

在引言写作时，为了引起读者的注意，采用一个问题或一系列问题开头不失为引言写作中一种十分有效的方法。当读者读到问题时，就会急切想知道问题的答案如何，给人一种悬念。例如在一篇论文中，其引言是用这样一个疑问句开头的"How much bureaucratic stupidity do we have to put up with?"当读者一读到这个问题就会产生"有多少官僚主义的愚蠢行为？是哪些？我们为什么会容忍"等一系列有待于从论文中寻求答案的问题。这样容易吸引读者，也便于读者了解论文的内容和进行检索时决定取舍。下面是一个以问题开头的引言的全文。

例2：

What are the chances of a nuclear war in the near future? How many Americans would survive a nuclear attack? Would such an attack make living conditions impossible for the survivors? These and similar questions are being asked by citizens' groups throughout this country and they debate the issue of arms control and nuclear disarmament.

例2是个以一系列问题开头的论文的引言。在引言一开始，作者用三个问题提出了美国全国人民关心核战的可能性，有多少人可以幸免于难，幸存者将来的生活条件是否存在。紧接问题之后,涉及了论文的主题——核裁军与军备控制,为论文对这一问题的讨论打下了基础。

要注意的是,所提的问题必须在论文中进行解答,同时所提的问题必须是文章中的内容。千万不要写出一些与文章主题关系不大或者根本没有关系的问题。

3. 引语开头法

利用引语开头是点明论文主题的一种有效方法。引语开头容易引起读者的注意。为了表示引语的权威性,可以在引语后增加出处,同时也为论文围绕这一主题展开讨论做好铺垫和提供强有力的依据。所以,在使用引语的时候,所引的话最好使用有一定权威性的著作,名人的名言或某一技术领域、专业权威人士的文字或语言。下面是一个引用著名医学家 Bernard Baruch 的话为开头的论文的引言。

例3:

There are no such things as incurables, said Bernard Baruch, "There are only things for which man has not found a cure." A lifetime in medical practice, education and research has convinced me, too, that we need not accept disease as an inescapable human destiny, despite our lack of information about many forms of human illness.

例 3 在一开头就引用了著名医学家 Bernard Baruch 一句关于治疗疾病的论断,他提出了应正确对待疾病。从中就可以看出这篇论文的主题是关于如何认识疾病和正确对待疾病的。

4. 定义开头法

用定义对文中所要论述的主题开头有利于作者表达,这样既可以点明论文的主题又可以对文中将要论述的中心词进行必要的解释。当然有时定义单用一句话解释不全,这时就要使用一段文字对其进行定义。例 4 就是一个使用一段文字定义作为引言的实例。

例 4:

A nurse is one who looks after patients. She helps the doctor, who cures sick persons by writing out prescription. And it is the nurse with whom the patients spend most of the time in a ward room.

例 4 是一个用定义"护士的职责和工作"为引言的典型实例。它首先解释了护士工作总的概念;然后叙述了护士的工作范围,并且用了一个非限制性定语从句非常巧妙地将护士的工作范围与医生的工作范围作了比较。从这个引言看,不难看出论文的主题将是叙述护士应怎样履行自己的职责。

5. 利用统计数字开头法

利用统计数字开头的引言是为了利用数字来说明某一问题。其目的是说明文章的主题的重要性,为在论文中对数据的采集方法、使用数据的目的进行论述前的准备工作。下面是一个以数据统计为开头的实例。

例 5:

The fact that less than 5% of the British population graduate from universities may seem surprising, especially when compared with American percentage of over 30%.

从例 5 中可以看出论文的主题是对英美两国人口中大学毕业生所占的比例进行分析,从中找出由于某些差异而形成的这一对比数字。

6. 假设情景开头法

假设一种情景是为了把主题的内容突出出来。在科技英语写作中,表示假设情景的最好句型是用"What if…"来引出，比如"What if there is no electric power?"那么其后的内容必定是围绕"如果没有电会是个什么样的情况"来做文章。这种用法的特点是给人一种悬念感，其作用与用问题开头写引言基本相同。但是，并不是除了"What if…"句型外就没有别的表达方式了，也可以把"imagine，think"这类词放在疑问句中来使用，表示假设。下面就是关于海伦·凯勒的文章的引言实例。

例6：
Can you imagine living your life without being able to see or hear anyone or anything? If that were not bad enough, imagine not being able to speak. You would feel totally isolated and cut off from the world, unable to communicate your feelings and reeds, and unable to share or understand those of others. That is how Helen Keller felt, before she met her teacher, Anne Mansfield Sullivan.

例6是以假设一种情况开始的引言。这个实例中的假设实际是一种对海伦·凯勒情况的真实描述。文章的主题可以从中推断出是海伦·凯勒在遇到她的老师沙利文之后生活起了变化的内容。真正的主题是隐含在这个引言的最后一句话中的。

引言写作关键要在"引"上下功夫。从上述介绍的6种方法中可以看出，引言除了它本身应讲述的内容外，主要还在于为正文的铺垫下功夫，为正文的内容起到一个引导的作用。在引言的最后要么明确指出下文，也就是论文的正文将要论述的内容（如例1），要么采用暗喻的办法（如例6）来作为引言的结束。

需要指出的是，如果引言是对某一自然学科的讨论，则人称往往使用第三人称单数，其时态根据需要使用一般现在时或过去时即可。如果引言是对某一人文社会科学问题的讨论，则可使用其他人称。

19.2.6 正文

正文是一篇论文的关键部分，所要论述的内容包括一些设想、实验情况、实验装置、获得的数据、证实的理论或新的设计方法等，均包含在这一部分。作者可通过定义、描写、说明、举例、实验、论证、比较对比、分析综合、推理判断等不同的研究过程来证实作者在引言中所提出的主题。

下面将对论文正文的写作方法逐一进行叙述。

1. 正文内容的顺序排列

英语科技论文中的正文一般有两种写法，一种是大体上按研究工作进程的时间顺序，依次叙述。在研究工作的实践中，原来是分不同层次进行的，经过多次循环，深化了对研究内容的认识。所以，在论文正文的写作中也要按照认识问题的先后，一个问题一个问题来写。安排问题的序列，要有其认识上的逻辑性。文章开始时，先对整个工作过程的层次应略有交代，然后，把一个问题作为一个层次内容，层层有实验结果，有小结，有导出下一层次工作的引子。最终有综合，有分析，有总的观点和结论性的结果。例如：有一项研究工作的第一层次，是在实验室设计实验，再现某一自然现象，使自然现象能在实验室中重复，从而抽象地得出现某现象的条件。第二层次是研究某现象存在的原因及其本质。第三层次是研究防止出现某现象的措施。第四层次是生产中的验证和改革生产工艺。如在正文中对上述研究成果进行描述时，可按每一循环中的感性认识和理性认识的螺旋上升，有层次地反映出来。每一层次有明确的小结,同时提出下一层次的研究方向。这样论文的正文就会层次分明,段落清楚,下笔也就自然多了。

另一种方法是按逻辑顺序排列,它与第一种时间排列顺序法不同。它是把研究工作全过程中多次由实践到理论的循环融合起来,提炼出典型的材料和观点,按认识由感性到理性的规律,逻辑地排列成章节。不是按试验工作的原有时间顺序,而是按照认识过程由低级向高级阶段的演变。一般首先介绍试验用的材料、试验设备、试验经过和试验结果。然后根据在试验过程中的步骤和实验结果,将数据和观察到的现象整理出来,加以综合,分别从实践上升到概念和判断。最后再进行必要的讨论,归结出结论,完成推理的阶段。

按逻辑排列的正文的写作方法有两种,一种是划分法,一种是分析法,分别叙述如下。

(1) 划分法

划分法是把事物划分为几个组成部分,分别予以处理。如一篇文章讲的是防止土壤侵蚀问题,可以根据内容将其分为4点,如:

Ⅰ. Erosion control in the East

Ⅱ. Erosion control in the South

Ⅲ. Erosion control in the North

Ⅳ. Erosion control in the West

然后再根据每一点的内容细分,如:

Ⅰ. Erosion control in the East

a. Erosion control in the past

b. Erosion control in the present

c. Erosion control in the future

由于这种划分非常方便,并且看起来也很清楚,有条理,有层次,十分合乎逻辑,因而在科技论文的正文写作中常常使用这种方法。

下面就是一个按照逻辑排列顺序的实例。

例1:

The amount of energy released from one atomic bomb is far greater than can be released at one time from ordinary sources of energy. This is because the atomic nucleus itself is burnt apart, and the nucleus of an atom has the most concentrated store of energy known. What is called atomic energy is actually nuclear energy.

There are two known ways to release the energy from the nucleus of an atom. One way is to split a nucleus of an atom into other nuclei of smaller mass. This method is called fission. The particles of the split atom produce heat through friction as they fly through the air. In addition, the reaction produces gamma rays and other tiny particles-neutrons-that can start fission in nearby atoms. The release of energy from a chain of these fission reactions in a pound of uranium-235 is greater than the release of energy from burning 2600000 pounds of coal.

The second known way to obtain energy from the nucleus is to bind nuclei together to form heavier nuclei. Some of the mass of the combining nuclei becomes energy. This process is called fusion. It is the source of the tremendous energy of the sun.

Much more energy can become available as the result of fusion than as the result of fission. Fusion is the principle of the hydrogen bomb. Scientists are now trying ways to control the tremendous energy released by fusion.

例1的第一段是概括性引言，因为它的最后一句话是中间段落的引子，所以也将其摘录。可以把第二段、第三段和第四段的文章内容按下述要点进行归纳。

Ⅱ Fission
  a. splitting a nucleus of an atom
  b. production of heat
  c. production of gamma rays and neutrons
  d. the results of fission
Ⅲ Binding nuclei together to form heavier nuclei
  a. production of energy-fussion
  b. an example-source of the sun's energy
Ⅳ Conclusion
  a. more energy from fusion
  b. the principle of fusion
  c. control fusion energy

同时，从例1中也可以清楚地看出，第一段和第三段的最后一句话都是为下一段的展开进行的必要的引导，从而使整篇文章之间的段落衔接更加紧密。

（2）分析法

分析法是由作者提出需要解决的问题，需要采取的措施或需要解释的现象。然后对其进行分析，直接找到解决这些问题的方法，或说明所采取的针对性措施的好处与不足，或解释清楚造成某些现象的原因；也可对几种可供选择的方法进行解释和对比，并证明为什么某种选择是最可接受的。

分析法常常用以下的方式表示：

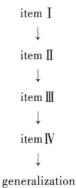

分析法实际上是归纳写作手法的体现。但是由于用其写作时对写作人员的文字功底要求较高，通常情况下若不是进行多项情况的比较，一般不宜采用。

2. 主题句的写作方法

主题句反映了一个中间段的中心思想，它也指出了这一段文字内容的发展方向与方式。主题句一般位于一段文字的句首，但也可放在中间或结尾。主题句所在的位置不同，有着不同的作用。段首是主题句最常见的位置，它可以使读者很容易了解到本段文字的内容。放在中间是因为这一段的前一部分内容要作为本段所叙述的内容的背景。放在最后则是采用了归纳法的写作方法，将该段文字的中心内容放在本段结尾时予以总结。通常，作为科技英语的段落写作往往把主题句放在段首，然后通过扩展细节和提供依据对主题句的内容进行论述。

需要指出的是，有人曾对英美论文作过统计，发现60%的主题句在段首，30%的主题句在段尾，10%的主题句在文章段落的中间。但是作为初学者来说，最好将主题句放在段首，以便使段落的内容按顺序展开。

下面就如何写好主题句和主题句写作时应注意的事项进行叙述。

（1）主题句的结构

要写好主题句，首先要弄清楚主题句的结构。主题句的结构通常包括两个部分，即：中心议题和控制部分。所谓中心议题就是该段文字所要涉及的人、事或问题；而且整个段落的文字内容都要围绕着这个人、这件事或这个问题展开。控制部分是决定该段文字的发展方向或扩展方法。它可能是一个词、短语或从句。控制部分暗示这段要由时序法（列举法）、空间顺序法、举例法、比较法、对比法或因果法等中的一种或几种方法混合在一起来发展。例如，有关篮球（basketball）这个主题，就可以写出下面5个主题句，每个主题句的控制部分都不一样，而且所使用的写作技巧也不一样（见画线部分）。

例2：

① Basketball has become more popular <u>within the past twenty years</u>. （时间）

② Basketball is a <u>physically demanding sport</u>. （列举）

③ Basketball and volleyball <u>have a great deal in common</u>. （比较）

④ Basketball is <u>less dangerous than football</u>. （对比）

⑤ Basketball is popular <u>for several reasons</u>. （原因）

从例2的5个实例中看，每个例句中的控制部分所表达的内容不一样，因此在各自例句后的段落中的文字内容也就自然不一样。例2中第1句的文字内容是叙述在近20年中篮球运动的普及情况。第2句指出篮球是一种大运动量的体育活动。第3句则是讲述篮球与排球之间的共同之处，其后的文字也就是要说清这两项球类运动之间的共性。第4句的控制部分将篮球与足球运动之间的危险性进行对比，那么该段的文字内容也必定要沿着这个方向进行发展。第5句是控制在几个原因上，作者就要在本段的文字中对这几个原因阐述清楚。

综上所述，主题句由中心议题与控制部分组成。中心议题是所讨论的对象，控制部分是关于所讨论的对象的具体内容，同时也限定了文字论述时所要运用的技巧。

（2）主题句的写作技巧与文体要求

主题句的写作质量会直接影响到整个文章的文字质量。因此，一篇科技英语文章的成功与否与主题句写作的质量有着至关重要的关系。

怎样写好主题句？在写好主题句时要注意哪些问题？主题句的文体要求又有什么样的特点？这些都是需要解决和注意的问题。

1）主题句必须有该段要讨论的中心议题和控制该段发展的范围，而控制范围常常用一个或几个关键词来表达。

例3：

① These experiments have been done in the newly-built laboratory.

② There exists air pollution in China.

③ Noise pollution is becoming more and more serious in this country.

在上述例子中，句①没有中心议题也没有该段的发展范围，无法继续发展下去。句②虽有中心议题但没有关键词来控制，使读者不知下一步要涉及的范围。句③既有中心议题又通过关键词来控制该段的内容，因此句③应该是一个较好的主题句。

2）主题句不能太泛、太一般化。这是由主题句的文体要求与写作特点所决定的，因为主题句太泛，整个段落就会显得太松散，从而对文章的内容难以控制。有的主题句太宽泛，给人的感觉就是这不是一个段落的主题句，而是一篇文章的题目或是一本书的书名。例如：

例 4：

Health and wealth, which is better?

此主题句作为一篇文章的题目是合适的，而作为一个段落的主题句则显得太大、太笼统。现将其作为一篇文章的题目，然后再根据其内容扩展为几个主题句。将题目细化。

例 5：

① Some people hold that wealth is more important than health.

② However, the majority of people believe that health is more important than wealth.

③ In my opinion, health is better than wealth.

在例 5 的主题句中，每个句子都可作为一段文字的主题内容。在句①中主议题所控制的扩展方向是"wealth is more important than health"，而且还限定了"some people"这个范畴。在句②中，表示了相反的观点，即"health is more important than wealth"，它所要求的扩展内容也只能符合这个方向。句③则是作为个人的观点提出一个结论性意见，其限制的内容就是："health is better than wealth"。例 5 的三个主题句中，语言结构简单，文字表达内容准确、具体，控制部分要求明确，达到了主题句的写作要求。

3）主题句写作也不能过于具体，过于确切。因为太具体、太确切的主题句无法在随后的段落文字中对其进行扩展，它道出的是个事实，也就是说它所叙述的是一个具体、翔实的内容，再也没有进一步发展与说明的必要。例如下面一组例句。

例 6：

① If there were no electric power, people would have to use fire to light the room in the evening.（太窄，无法展开）

② If there were no electric power, things would be quite different.（清楚，明确，恰当）

从例 6 来看，句①已经道出了事实，而且还提供了具体的细节，应该说对其是无法进行扩展的。而句②的主题句使用了"things would be quite different"这个发展方向，使作者可使用列举的写作方法对"quite different"进行必要和具体的描述。

4）因为主题句要清楚、直接地表达一个中间段的中心思想，所以主题句的句型常常是简单句或简洁的句子。试比较下列一组实例。

例 7：

① In the United States, the system of forced labour, which was known as slavery, lasted almost 250 years.（冗长）

② Slavery lasted almost 250 years in the United States.（简明，清楚）

在例 7 中，第①句话使用了非限制性定语从句，而且还有一些不必要的重复。在第②句话只是一个简单句,句中首先点明了 slavery 这个主题,然后用时间来表明要在下述文字中应扩展的方向,文字表述得准确、清楚、简明,是一个质量较高的主题句。

5) 在主题句中也可以使用疑问句,特别是使用特殊疑问句。使用疑问句的好处是易引起读者的兴趣,帮助读者迅速确定文字的扩展方向和内容。

例 8:

① What is meant by scientific attitude?

② What are the possible reasons for the decline of car accidents in big cities?

在例 8 的两个例句中,可以清楚地看出:第①句是要对"scientific attitude"一词做出解释,在这一段后的文字中要求作者围绕这一主题展开讨论。第②句的主题句后的文字,则要求作者对"possible reasons"进行分析和解释。但是,一般来说,在段落的主题句中若能不采用疑问句方式应尽量不要采用疑问句来作为一段文字的主题句。这是因为科技英语文章的文体中用疑问句作主题句有时会显得笼统,有时口气显得比较强烈且往往会带有个人的感情色彩。

3. 扩展句的写作方法

在写好一个段落的主题句后,就应按照主题句中的控制部分来扩展该段落的内容,以完成整个段落的写作。也就是说,要根据主题句中控制部分的内容对其进行扩展。具体地说,要对控制部分的内容要求进行讨论、说明、支撑或者证明等,那么,在扩展句中所要使用的是事实、理由、原因、说明、图表、比较、对比、定义等方式来阐述主题句中控制部分的思想内容。所以,一段写作质量较高的文字必须遵循以下三个原则。

(1) 一致性

一段好的文字最重要的特点是一致性,在写作一个段落时,只能讨论问题的一个方面或一个内容,这就叫作文章的一致性。所以,在一段文字中,所使用的事例、所叙述的事实以及所解释的理由等,都必须与所叙述的主题有关。如果所要使用的材料与这段文字的主题没有直接的关系,那么这一段文字就不会是成功的写作结果了。

(2) 连贯性

连贯性就是要在一段文字中将所有的句子清楚地、有逻辑地并按先后顺序联系在一起,共同说明这一段的主题。在表达连贯性很强的段落中,每一句话都应自然地从前一句中繁衍而出,对该段的中心思想进行扩展。理想地说,这样的一段文字应具有一种自然流畅感,读者也就轻而易举地发现所有这一段中的句子从逻辑上是联系在一起的。所以在这样的一段文字中,可采用时间顺序法、空间描述法、比较对比法以及按重要性来排列一段文字中各句的顺序等写作手法。

为了把连贯性表达清楚,可以使用的写作手法有很多,如:时间顺序法、空间顺序法、演绎法、定义法、比较法、对比法、归纳法、高潮顺序法和混合顺序法等。但是不管使用何种方法都必须首先考虑到主题句中控制部分的要求,然后再使用不同类型的连词、代词(用以指代前句中提到的名词,因为它们可以把一段中的句子与句子连接起来),以及使用必要的重复,如重复某些关键的词汇和词组,或使用同义词或词组来表示重复。这样做能使读者记住前面的内容,同时把各句也都联系起来了。

(3) 完整性

如果一个段落没有完整地或充分地说明主题句所要求叙述的内容,那么这个段落就不是一个意义完整的段落。所谓完整性就是要将主题句中要求表示的内容用一致性和连贯性的方式完整地表达清楚，而不是用字数多长来进行衡量。如果主题句限制的内容较小，所写的段落也就应较短。如果主题句限制的内容较大，那么该段的文字也就会较长。所以，完整性应首先按照主题句所要求表示的内容来提供具体、翔实而具有说服力的事实。其次就是要运用好各种写作技巧，使文章段落结构严谨。

（4）写好扩展句的文体要求及注意事项

在科技英语写作中，如果说主题句是一个人的头部，那么扩展句就应该是一个人的躯干。扩展句是对主题句中的主题部分按控制部分的要求，将其清楚、完整、准确地描述出来。因此，在写扩展句时，除了要注意一致性、连贯性和完整性外，在写作文体和其他方面还要注意以下的特点与要求。

1）文字的逻辑性与条理性。在文体方面除了一致性、连贯性和完整性外，还要注意文字的逻辑性与条理性。要做到逻辑性强和条理清楚，就要使用好转接词。因为转接词用来表明文章的发展线索，它们可能是连词、代词或是被重复的单词、词组。

2）避免重复。在英语文章中，无论是在科技文章或是在文学作品中，都要通过使用同义词、同义词词组或者代词来取代或指代前文中可以取代或指代的词汇与词组。

3）尽量不使用意义模棱两可的修饰语。科技英语文章文体特征之一就是准确。如果使用诸如 almost，nearly，about 等这类词汇就往往达不到准确表达一个概念、说明一个过程、描述一个物体的基本要求。

4）正确使用人称。论文常用的人称主要体现在使用人称代词上。在人称代词的使用上有两种主张，一种是传统式的主张，认为科技文章应侧重叙事和推理，读者重视的是论文的内容和观点，感兴趣的是作者的发现，不是作者本人。因此要避免使用第一、第二人称代词，尤其不能使用第一人称单数 I，认为 I 是主观的象征，与科学是不相称的，所以就导致了大量使用被动语态。另一种主张则强调文章应自然，直截了当，要使用第一、第二人称代词。

5）正确使用时态。在科技英语写作中，论文的作者要向读者表述研究过程中各项事实、观点产生的时间，所以正确使用时态就成为一个十分值得注意的问题。常见的方法有以下几种。

① 表示研究目的的动词时态一般要使用过去时，这是因为研究的目的是在着手研究时确定的，见例 1。

例 1：

The aim of this study was to solve the calculating method of this problem.

② 表示结论的动词时态一般使用现在时态或现在完成时态，而且以使用现在时为主，除非在强调已获得的成果时方可使用现在完成时。

例 2：

We conclude that the principles of the test system allow increased safety and accuracy in hospital drug handling.

③ 表示研究过程中动作或状态的动词时态一般使用过去时，这是因为研究工作是在撰写论文之前进行的。所以，当在论文中叙述作者做了哪些研究工作、研究对象的情况、研究工作过程中出现了什么现象、得出什么结果时，所描述的都是过去做过的事，而只能使用过去时态。但有时也会强调某一行为或状态已完成或持续到撰写论文时，也会使用现在完成时。

例 3：

The results of treatment of early gastric carcinoma were analysed in 65 patients.

④ 说明图表的时态一般使用现在时。

例4：

Diagrams showing yields are shown in Figure 3. The second column of Table 2 represents the dry weight of tops.

4. 结尾句

结尾句放在一段的末尾，总结这一段的要点。结尾句通常与主题句一样包括了该段文字的控制思路。当使用到主题句的控制思路时，在结尾句中所使用的词汇要与主题句中控制部分所使用的词汇有所不同。要写结尾句，就要抓住主题句中的关键词，或回答主题句中所暗示的提问，有时要总结一段的要点，可能会用到 to sum up, in short, in conclusion 等词组或词汇。

19.2.7 结论

1. 结论段的文体特点

结论段中的文字是要将讨论予以结束，如果有必要，还要对其进行总结。英语科技论文的结尾不外乎起三种作用：①使读者对于文章的内容得到一个全面的清晰而明确的印象；②使读者能够接受文章内容的启发，自己进一步去发现问题、思考问题；③给读者以强有力的感应，促使读者对讨论的问题或与之相关的问题进行更深入的研究和探讨。

因此，可以说结尾是文章逻辑发展的必然和自然结果。在结尾段中要反映经过分析、综合、推理、判断、比较、实验等研究过程所形成的总的观念。在结尾段中要把文章各部分综合联系起来，突出文章的中心思想，构成文章不可缺少的一部分。另外，还要注意开题与结尾前后呼应，做到首尾一致。

下面是常见的几个科研论文的结尾的实例。

例1：

Conclusion

These two new windmill designs will, of course, have an even more important impact on the energy industry in the future. Although these two designs may never be used on a large scale, their impact may be felt through influencing the design of even more efficient models in the years ahead.

在例1中，首先使用了小标题 Conclusion，这是科研论文常用来概括性地表明要研究讨论的内容的一种表达方式。本段除了对文章的主题进行概括性总结外，还在最后一句话中对所研究的工作在未来的可能性进行了必要的预测。

例2则是另一种文体的结尾段，它是回答某一问题的文章结尾。文章的标题是："Why the Sky Looks Blue"

例2：

Conclusion

Thus, the different constituents of sunlight are treated in different ways as they struggle through the earth's atmosphere. A wave of blue light may be scattered by a dust particle, and turned out of its course, and so on. Until finally it enters our eyes by a path as zigzag as that of a flash of lightening. Consequently, the blue waves of the sunlight enter our eyes from all directions. And that is why the sky looks blue.

例2的最后一句是对全文所讨论的问题的答案和总结。

例3是叙述一个实验并对其作用加以讨论的文章的结尾，其特点是仅用一句话就指出这个实验的优点，以高度概括的手法总结全文的内容。

例3：

Conclusion

In summary, the exercise introduces the students to the application of drug analysis in whole blood and serum without burdening him with excessive theoretical and technical considerations.

总而言之，由于英语科技论文的特点、内容与结构的不同，文章的结尾也不同。可以说因为论文的形式多样，所以文章的结尾也就千姿百态。在这里，仅举了最常见的几种来说明结尾常包含的内容和运用的形式，以起到写论文结尾段落时的引导作用。

2. 写结论段时应注意的几个问题

写英文的科技论文时必须考虑到英美文化的特点以及英汉文化的差异，在写结尾段时要注意避免下述几个问题。

1) 不必要的谦虚。科技论文的特点是实事求是，研究的成果是什么就是什么，既不要吹嘘，也不要过分谦虚，这是英美文化中最忌讳的事。如在文章的结尾段中就不应出现主观性很强的表示。

例1：

There must be many mistakes and shortcomings in my paper. I hope you will criticize my paper.

这样的语句一旦出现在结论段中，无论论文的正文所讨论的问题的意义有多大，所取得的研究成果何等重要，在别人的眼里也是不值得一看的一篇论文。

2) 在结尾时不要削弱或偏离文章的主题。如在一篇叙述计算机作用和用途的文章的结尾时，突然冒出了例2这样的句子，就会使人感到与文章的主题相违背，而成了另一个议题。

例2：

Of course, computer has its harmful social effects.

显然，应把例2作为另一议题的主题句来讨论，而不是对计算机的作用和用途进行总结。这本身也就违背了文章内容的一致性与连贯性的要求。

3) 没有必要的强调或使用不当的语句来结束文章。例3就是这样的一个实例。

例3：

This problem deserves the serious attention of every scientist in this field.

例3这句话作为文章的结尾显然多余。

4) 避免使用口语化的表达方式。这是英语科技文章的文体要求所决定的。如果在结论段的开始使用了例4这样的表达方式，就会使人感到文章成了一个演讲的书面稿。

例4：

So I want to close my paper by summarizing the above discussion.

3. 在结论段中常用的表达方式与句型

在结尾段中常用的表达方式有：

In conclusion/summary...

在结尾段中常用的词汇与词组有：

Thus, hence, therefore, put things together 等。

表示结论的常用句型有：

The course program/the experiment described above is…

It is clear from the foregoing discussion that…

This demonstration/illustration/comparison gives…

19.2.8 致谢

由于科学研究工作常常不是一个人或几个人的力量所能完成的，有时需要有关单位和个人的指导和支援。在论文后"致谢"中对给予过帮助的单位和个人表示感谢，以说明其在该项工作中的贡献和责任。

"致谢"通常由以下内容组成：

1）对为研究工作提供方便和帮助的实验室或个人表示感谢。尤其是对为研究提供专门实验设备或其他材料的人表示感谢。

2）对经济上提供支持的人表示感谢。例如通过赠送、签订合同或研究基金等形式为论文作者提供课题基金计划以外的资金援助的个人或机构，应对其表示感谢。

3）为论文的内容提出过意见或建议，或者进行过其他形式的帮助，或者审阅过论文的人，应对他们的工作表示感谢。在致谢中应如实地指出其具体内容，说明他们所起的作用。

4）对打字员、绘图员以及其他工作人员在论文中所做的工作，可以在致谢中提及，但没有必要专门对他们的工作致谢，因为这是他们的正常工作。

例1：

Acknowledgement

The author is indebted to Dr. Dale for suggesting the use of Ka kutani's theorem to simplify the proof and to the A. E. C. for financial support.

例2：

Acknowledgement

The author wishes to express his most sincere thanks to Prof. Gao Zhengheng, who read the manuscript carefully and gave valuable advice. Tremendous thanks are owed to Mr. Wu Xuehan, Wang Conghui and Jiang Yongcai for helping us with elemental analysis, mass spectral analysis and emission spectra recording. Also, the author is very grateful to Shanghai First and Twelfth Dyes Factories for offering some samples.

以上是两个致谢的实例。在两个实例中，作者均对为论文提供过帮助的个人或单位以及对他们所做的工作表示了谢意。

1）在致谢时第一句话常用的句型有：

The author wished to express his/her most sincere thanks to…

The author is indebted to… for…

The author is very grateful to… for…

We would like to thank… for…

The author wishes to acknowledge… for…

We are gratefully to recognize…

2）致谢时常用的词汇与词组有：

to be owed to，recognize，acknowledge，contribute to，help，contribution 等。

3）致谢的文体一般采用例 1 与例 2 的格式。但有时也先叙述被致谢的个人与单位在论文中所给予的帮助和所起到的作用，然后再致谢。使用后一种致谢文体，往往是因为被致谢的个人与单位较多。为了表达清楚起见，就采用这种形式。

19.2.9 参考文献

在科技论文之后列出参考文献是为了反映作者严肃的科学态度和研究工作的依据，也是维护知识产权的需要。所以，参考文献也是论文写作的一个部分。

根据英语科技论文写作的规定，凡是引用其他作者的文章、观点或研究成果，都应标明，并在参考文献栏中说明出处。参考文献可以引用杂志论文、专利、毕业论文、专著及未出版的著作。引用要完整、清楚，以便读者需要阅读该文献时可以找到。现将国际标准组织对参考文献的各种要求分别叙述如下。

1. 表示参考文献的 2 种顺序

根据国际标准组织的规定，参考文献顺序的排列有两种形式。一种是按姓的首字母在字母表中的顺序进行排列；一种是按在正文中引用的顺序排列。

（1）按姓的首字母顺序排列

这种排列方式是一种传统的排列方式。在这种排列方式中，参考文献的格式与脚注的形式基本相同，不同的是将作者的姓放在第一个词的位置，并依照其字母的顺序来排列在参考文献中的顺序。如果一作者有两种著作，作者姓名在第二次和以后可以用一条半英寸长的实线来代替。对没有署名的参考文献可根据文章和书目的第一个词汇的字母顺序排列。如果是一个作者有两本著作或者论文，则按著作和论文的第一个单词的首字母的顺序进行排列。第二行应对齐上行的第五个字母。在列出参考文献时，首先要标出小标题 References。这个小标题既可放在中央，也可放在左对齐位置。

需要说明的是：①作者的姓用全称，名字中只用第一个字母；②一篇论文有 3 个以上的作者时，可只使用第一位作者的姓名，其后加上 et al；②杂志名称可按 ISO 833—1974 的要求缩写。

（2）按数字标号列出的参考文献

在不使用脚注的情况下，可按数字顺序来排列参考文献。如果使用这种方法，所列出的参考文献应与文章中标注的顺序一致。这种排列方式与按姓的首字母排列顺序的方法一致，唯一不同的是姓名的书写方式没有顺序的变化。第二行只需同第一行的文字的首字母对齐即可。

2. 引用不同种类的参考文献的表示方法

由于参考文献可包括杂志、著作、专利、未出版的作品等。在参考文献中的格式也有所不同。下面我们以数字排序的方法对不同参考文献的表示方法进行分别叙述。

（1）引用杂志论文

引用杂志论文的顺序是：①作者姓名（其中第一、第二个名字缩写）；②论文标题；③杂志名称（用斜体字，可按 ISO 833—1974 缩写）；④卷号（可省略掉 Vol. 的字样）；⑤期号；⑥引用论文的首页与末页。

（2）引用专利

引用专利的格式是：①发明者姓名；②国别与专利号；③专利使用年代；④拥有专利的机构；⑤可以找到专利摘要的摘要杂志。

需要说明的是，引用专利时，总要与它的摘要（二次文献）在一起。

（3）引用毕业论文

引用毕业论文的格式是：①作者姓名；②何种学位论文；③发表论文的大学；④大学所在城市；⑤论文完成的年代。

（4）引用书籍

引用的书籍分为两种类型：一种是专著，另一种是合著。但是其格式基本相同，包括①作者姓名；②书名（用斜体字打印）；③版次；④出版社或公司；⑤出版地与年代；⑥卷号；⑦页码。要注意的是，书名不可缩写。多人合作的著作，作者在三人以上时只使用第一著者的名字即可。

（5）引用未出版的作品

虽然有的作品尚未出版或发表，但也可对其中的观点和内容进行引用。引用的方法比较简单。

1）引用没有出版而且也没被决定采用的作品，其表达方式如下：

J. Smith, unpublished work.

2）引用没有出版但已被杂志或出版社接受决定即将发表的作品时，其表达方式如下：

J. Smith, Spectrochim. *Acta*, in press.

（6）引用对话作为参考文献

引用对话作为参考文献的格式如下：

B. Miller, Personal communication.

参 考 文 献

［1］ American Society of Civil Engineers. ASCE/SEI 7—2005 Minimum Design Loads for Buildings and Other Structures ［S］. Reston, VA: ASCE, 2005.

［2］ International Code Council. International Building Code ［S］. Washington, DC: ICC, 2006.

［3］ American Association of State Highway and Transportation Officials (AASHTO). AASHTO LRFD Bridge Design Specifications ［S］. Washington, DC: AASHTO, 2008.

［4］ Building Seismic Safety Council. NEHRP Recommended Provisions for Seismic Regulations for New Buildings and Other Structures ［M］. Washington, DC: Building Seismic Safety Council, 2001.

［5］ Macgregor J G, Mirza S A, Ellingwood B R. Statistical analysis of resistance of reinforced and prestressed concrete members ［J］. Journal of the American Concrete Institute, 1983, 80 (3): 167-176.

［6］ Macgregor J G. Load and resistance factors for concrete design ［J］. Journal of the American Concrete Institute, 1983, 80 (4): 279-287.

［7］ Macgregor J G. Safety and limit states design for reinforced concrete ［J］. Canadian Journal of Civil Engineering, 1976, 3 (3): 484-513.

［8］ Winter G. Safety and Serviceability Provisions of the ACI Building Code ［J］. ACI Special Publication, 1979, 59.

［9］ Nowak A S, Nowak S, Szerszen M M. Calibration of design code for buildings (ACI 318): Part 1 - Statistical models for resistance ［J］. Aci Structural Journal, 2003, 100 (3): 377-382.

［10］ Szerszen M M, Nowak A S. Calibration of design code for buildings (ACI 318): Part 2 - Reliability analysis and resistance factors ［J］. Aci Structural Journal, 2003, 100 (3): 383-391.

［11］ American Concrete Institute. Building Code Requirements for Structural Concrete and Commentary (ACI 318-08) ［S］. Farmington Hills, Ml: ACI, 2008.

［12］ Gerard P. ICE Manual of Bridge Engineering ［M］. London: Thomas Telford Publishing, 2000.

［13］ Salter R J. Highway design and construction ［M］. 2nd ed. London: Macmillan Education UK, 1988.

［14］ Slinn M, Matthews P, Guest P. Traffic Engineering Design ［M］. 2nd ed. Oxford: Elsevier, 2005.

［15］ Hager I, Golonka A, Putanowicz R. 3D Printing of Buildings and Building Components as the Future of Sustainable Construction ［J］. Procedia Engineering, 2016, 151: 292-299.

［16］ Xu J, Ding L, Love P E D. Digital reproduction of historical building ornamental components: From 3D scanning to 3D printing ［J］. Automation in Construction, 2017, 76: 85-96.

［17］ Buckley P J, Dunning J H, Pearce R D. The Influence of Firm Size, Industry, Nationality, and Degree of Multinationality on the Growth and Profitability of the World's Largest Firms, 1962-1972 ［J］. Weltwirtschaftliches Archiv, 1978, 114 (2): 243-257.

［18］ Daniels J D, Bracker J. Profit Performance: Do Foreign Operations Make a Difference? ［J］. Management International Review, 1989, 29 (1): 46-56.

［19］ Department of Foreign Economic Cooperation. Brief statistics of contracted projects, labor services and design and consultancy services in 2000 ［R］. Beijing: Ministry of Foreign Trade and Economic Cooperation, 2000.

［20］ Department of Foreign Economic Cooperation. Brief statistics of foreign economic cooperation 2001 ［R］. Beijing: Ministry of Foreign Trade and Economic Cooperation, 2001.

［21］ Zhang S, Chen Q, Jin J. An Empirical Study on the Region-Based Housing Industrialization Grading System in Chinese Rural Areas ［C］ // International Conference on Construction and Real Estate Management. 2015: 435-444.

［22］ Huang Z, Li G, Song Z. Development Path Exploration of China's Housing Industry from the Perspective of Market Analysis ［C］ // International Conference on Construction and Real Estate Management. 2015: 877-884.

［23］ Badir Y F, Kadir M R A, Hashim A H. Industrialized Building Systems Construction in Malaysia ［J］. Journal of Architectural Engineering, 2002, 8 (1): 19-23.

［24］ Chen Y, Okudan G E, Riley D R. Decision support for construction method selection in concrete buildings: Prefabrication adoption and optimization [J]. Automation in Construction, 2010, 19 (6): 665-675.

[25] Su L, Ahmadi G, Tadjbakhsh I G. Comparative Study of Base Isolation Systems [J]. Journal of Engineering Mechanics, 1989, 115 (9): 1976-1992.

[26] Lee S, Yu J, Jeong D. BIM Acceptance Model in Construction Organizations [J]. Journal of Management in Engineering, 2013, 31 (3): 04014048.

[27] Azhar S. Building Information Modeling (BIM): Trends, Benefits, Risks and Challenges for the AEC Industry [J]. Leadership and management in Engineering, 2011, 11 (3): 241-252.

[28] Azhar S, Hein M, Sketo B. Building Information Modeling (BIM): Benefits, Risks and Challenges [J]. Town Planning Review, 2004, 75 (2): 231-255.

[29] Horvat I E. The future of underground infrastructure in Holland [J]. Tunnelling & Underground Space Technology, 1996, 11 (2): 258-260.

[30] Zuo D Q, Gu Z X, Wang W. Barrages, spillways and control works [M] //Handbook of hydraulic structure design. Beijing: Water Resources and Electric Power Press of China, 1987.

[31] Lin Q X, Liu Y M, Tham L G, et al. Time-dependent strength degradation of granite [J]. International Journal of Rock Mechanics & Mining Sciences, 2009, 46 (7): 1103-1114.

[32] Ministry of Water Resources of the People's Republic of China SL 60—1994. Technical criterion on Earth-Rockfill Dam safety monitoring [M]. Beijing: China Water Power Press, 1994.

[33] Raphael J M, Carlson R W. Measurement of structural action in dams [M]. 3rd ed. Berkeley, Calif.: Gillick, 1956.

[34] Applied Technology Council. Proceedtngs of ATC-17-1 Seminar on Seismic Isolation, Passive energy dissipation and active control [C]. Redwood City: CA, 1993.

[35] Soong T T, Dargush G F. Passive energy dissipation systems in structural engineering [M]. London: Wiley, 1997.

[36] Constantinou M C, Soong T T, Dargush G F. Passive Energy Dissipation Systems for Structural Design and Retrofit [J]. Center for Earthquake Engineering Research State University of New, 1998.

[37] Soong T T. Active structural control: theory and practice [M]. London: Longman, 1990.

[38] Suhardjo J, Jr B F S, Sain M K. Feedback-feedforward control of structures under seismic excitation [J]. Structural Safety, 1990, 8 (1-4): 69-89.

[39] Housner G W, Soong T T, Masri S F. Second Generation of Active Structural Control in Civil Engineering [J]. Computer-Aided Civil and Infrastructure Engineering, 2010, 11 (5): 289-296.

[40] Housner G W, BergmanV T, Caughey K, et al. Structural control: past, present, and future [J]. Journal of Engineering Mechanics, 1997, 123 (9): 897-971.

[41] Kobori T. Future Direction on Research and Development of Seismic-Response-Controlled Structures [J]. Computer-Aided Civil and Infrastructure Engineering, 1996, 11 (5): 297-304.

[42] Dyke S J, Jr B F S, Quast P, et al. The Role of Control-Structure Interaction in Protective System Design [J]. Journal of Engineering Mechanics, 1995, 121 (2): 322-338.

[43] Aiken I D, Nims D K, Kelly J M. Comparative study of four passive energy dissipation systems [J]. Bulletin of the New Zealand Society for Earthquake Engineering, 1992, 25 (3): 175-192.

[44] Skinner R I, Tyler R G, Heine A J, et al. Hysteretic dampers for the protection of structures from earthquakes [J]. Bulletin of the New Zealand National Society for Earthquake Engineering, 1980, 13 (1): 22-36.

[45] Fujita T. Seismic isolation and response control for nuclear and non-nuclear structures: Special Issue for the Exhibition of 11th Int. Conf. on SMIRT, Tokyo, 1991 [C].

[46] Martinez-Romero E. Experiences on the use of supplemental energy dissipators on building structures [J]. Earthquake Spectra, 1993, 9 (3): 581-624.

[47] Perry C L, Fierro E A, Sedarat H, et al. Seismic upgrade in San Francisco using energy dissipation devices [J]. Earthquake Spectra, 1993, 9 (3): 559-579.

[48] Wada A, Huang Y H, Iwata M. Passive damping technology for buildings in Japan [J]. Progress in Structural Engineering and Materials, 2000, 2: 335-350.

[49] Clark P W, Aiken I D, Tajirian F, et al. Design procedures for buildings incorporating hysteretic damping devices: Proc. Int. Post-Smirt Conf. Seminar on Seismic Isolation, Passive Energy Dissipation and Active Control of Vibrations of Structures, Cheju, 1999 [C].

[50] Pall A S, Marsh C. Response of friction damped braced frames [J]. Journal of the Structural Division, 1982, 108 (6): 1-5.

[51] Grigorian C E, Yang T S, Popov E P. Slotted bolted connection energy dissipators [J]. Earthquake Spectra, 1993, 9 (3): 491-504.

[52] Pall A S, Pall R. Friction-dampers used for seismic control of new and existing building in Canada: Proc. ATC 17-1 Seminar on Isolation, Energy Dissipation and Active Control, San Francisco, 1993 [C].

[53] Chang K C, Shen K L, Soong T T, et al. Seismic retrofit of a concrete frame with added viscoelastic dampers: 5th Nat. Conf. on Earthquake Engineering, Chicago, 1994 [C].

[54] Shen K L, Soong T T, Chang K C, et al. Seismic behaviour of reinforced concrete frame with added viscoelastic dampers [J]. Engineering Structures, 1995, 17 (5): 372-380.

[55] Hao D S, Chang K C, Soong T T, et al. Full-Scale Viscoelastically Damped Steel Frame [J]. Journal of Structural Engineering, 1995, 121 (10): 1443-1447.

[56] Crosby P, Kelly J, Singh J P. Utilizing Visco-Elastic Dampers in the Seismic Retrofit of a Thirteen Story Steel Framed Building [C] // Structures Congress XII. ASCE, 2015: 1286-1291.

[57] Soong T T, Reinhorn A M, Nielsen E J, et al. Seismic upgrade of a reinforced concrete building using viscoelastic dampers: Proc. World Structural Engineers Congress, San Francisco, 1998 [C].

[58] Miranda E, Alonso J, Lai M L. Performance-based design of a building in Mexico City using viscoelastic dampers: Proc. Sixth US National Conference on Earthquake Engineering, Seattle, 1998 [C].

[59] Constantinou M C, Symans M D. Experimental study of seismic response of buildings with supplemental fluid dampers [J]. Structural Design of Tall Buildings, 2010, 2 (2): 93-132.

[60] Fujino Y. Recent research and developments on control of bridges under wind and traffic excitations in Japan: Proc. Int. Workshop on Struct. Control, Honolulu, 1994 [C].

[61] Spencer B F J, Sain M K. Controlling buildings: a new frontier in feedback [J]. Control Systems IEEE, 1997, 17 (6): 19-35.

[62] Soong T T, Reinhorn A M, Aizawa S, et al. Recent structural applications of active control technology [J]. Journal of Structural Control, 2010, 1 (1-2): 1-21.

[63] Yamazaki S, Nagata N, Abiru H. Tuned active dampers installed in the Minato Mirai (MM) 21 landmark tower in Yokohama [J]. Journal of Wind Engineering & Industrial Aerodynamics, 1992, 43 (1-3): 1937-1948.

[64] Reinhorn A M, Soong T T, Helgeson R J, et al. Analysis, design and implementation of an active mass damper for a communication tower: Proc. 2nd World Conf. on Struct. Control, Kyoto, 1998 [C].

[65] Sack R L, Patten W. Semi-active hydraulic structural control: Proc. Int. Workshop on Struct. Control, Honolulu, 1993 [C].

[66] Patten W N. The I-35 Walnut Creek Bridge: an intelligent highway bridge via semi-active structural control: Proc. 2nd World Conf. on Struct. Control, Kyoto, 1998 [C].

[67] Patten W, Sun J, Li G, et al. Field test of an intelligent stiffener for bridges at the I-35 Walnut Creek Bridge [J]. Earthquake Engineering & Structural Dynamics, 1999, 28 (2): 109-126.

[68] Kuehn J, Song G, Sun J. Experimental verification of a non-protruding intelligent stiffener for bridges (ISB): Proc. Int. Post-SMIRT Conf. Seminar on Seismic Isolation, Passive Energy, Dissipation and Active Control of Vibrations of Structures, Cheju, 1999 [C].

[69] Kobori T, Takahashi M, Nasu T, et al. Seismic response controlled structure with Active Variable Stiffness system [J]. Earthquake Engineering & Structural Dynamics, 2010, 22 (11): 925-941.

[70] Kamagata S, Kobori T. Autonomous adaptive control of active variable stiffness systems for seismic ground motion: Proc. 1st World Conf. on Struct. Control, Pasadena, 1994 [C].

[71] Kobori T. Mission and perspective towards future structural control research: Proc. of 2nd World Conf. in Struct. Control, Kyoto, 1998 [C].

[72] Kurata N, Kobori T, Takahashi M, et al. Actual seismic response controlled building with semi-active damper system [J]. Earthquake Engineering & Structural Dynamics, 1999, 28 (1): 1427-1448.

[73] Shames I H, Cozzarelli F A. Elastic and inelastic stress analysis [M]. Prentice Hall: Englewood Cliffs, 1992.

[74] Rabinow J. The magnetic fluid clutch [J]. Electrical Engineers Journal of the Institution, 1951, 67 (12): 1167.

[75] Weiss K D, Carlson J D. A growing attraction to magnetic fluids [J]. Machine Design, 1994, 66 (15): 61-64.

[76] Carlson J D, Catanzarite D M, Clair K A. Commercial magneto- rheological fluid devices: Proc. 5th Int. Conf. ER Fluids, Sheffield, 1995 [C].

[77] Jr Spencer B F, Dyke S J, Sain M, et al. Idealized model of a magnetorheological damper: Proc. 12th Conf. on Analysis and Computation, Chicago, 1996 [C].

[78] Jr B F S, Dyke S J, Sain M K, et al. Phenomenological Model of a Magnetorheological Damper [J]. Journal of Engineering Mechanics, 1996, 123 (3): 230-238.

[79] Dyke S J, Jr B F S, Sain M K, et al. Experimental verification of semi-active structural control strategies using acceleration feedback: Proc. 3rd Int. Conf. on Motion and Vibration Control, Chiba, 1996 [C].

[80] Dyke S J, Jr B F S, Sain M K, et al. Modeling and control of magnetorheological dampers for seismic response reduction [J]. Smart Materials & Structures, 1996, 5 (5): 565.

[81] Dyke S J, Jr B F S, Sain M K, et al. An experimental study of MR dampers for seismic protection [J]. Smart Materials & Structures, 1998, 7 (5): 693-703.

[82] Nilson A H, Darwin D. Design of concrete structures [M]. 14th ed. New York: The McGraw-Hill Companies, Inc, 2010.

[83] Salmon C G, Johnson J E, Malhas F A. Steel structures design and behavior [M]. 4th ed. New York: Harper & Row, 1996.

[84] Murthy V N S. Advanced Foundation Engineering [M]. India: Satish Kumar Jain for CBS Publishers & Distributors, 2007.

[85] Lyang J, Lee D, Kung J. Bridge Engineering Handbook [M] Boca Raton: CRC Press LLC, 2000.

[86] Hashash Y M A, Hook J J, Schmidt B, et al. Seismic design and analysis of underground structures [J]. Tunnelling & Underground Space Technology Incorporating Trenchless Technology Research, 2001, 16 (4): 247-293.

[87] Chen S H. Hydraulic Structures [M]. Berlin/Heidelberg: Springer, 2015.

[88] Naeim F, Kelly J M. Design of Seismic Isolated Structures [M]. Hoboken: John Wiley & sons, Inc., 1999.

[89] Kelly T E. Base isolation of structures, Design guidelines [M]. Wellington: Holmes Consulting Group Ltd, 2001.

[90] Soong T T, Spencer B F. Supplemental energy dissipation: state-of-the-art and state-of-the-practice [J]. Engineering Structures, 2002, 24 (3): 243-259.

[91] Low S P, Jiang H B. Internationalization of Chinese Construction Enterprises [J]. Journal of Construction Engineering & Management, 2003, 129 (6): 589-598.

[92] Erik W L, Clifford F G. Project management: the managerial process [M]. 6th ed. New York: McGraw-Hill Education, 2014.

[93] Viswanath H R, Tolloczko J, Clarke J N. Multi-purpose High-rise Towers and Tall Buildings [M]. Lodon: Taylor & Francis, 1998.

[94] 刘坚, 周东华, 王文达. 钢与混凝土组合结构设计原理 [M]. 北京: 科学出版社, 2005.

[95] 郭向荣, 陈政清. 土木工程专业英语 [M]. 北京: 中国铁道出版社, 2001.

[96] 武秀丽. 土木工程专业英语 [M]. 北京: 中国铁道出版社, 2000.

[97] 赫江华, 周现伟, 赫丽, 等. 钢-混凝土组合梁的综述 [J]. 水利与建筑工程学报, 2011, 9 (2): 160-164.

[98] 秦卫红. 土木工程专业英语 [M]. 武汉: 华中科技大学出版社, 2012.

[99] 王建武, 李民权, 曹小珊. 科技英语写作——写作技巧、范文 [M]. 西安: 西北工业大学出版社, 2000.

[100] 秦荻辉. 科技英语写作教程 [M]. 西安: 西安电子科技大学出版社, 2001.